HRW
TECHNOLOGY
HANDBOOK

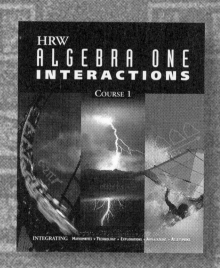

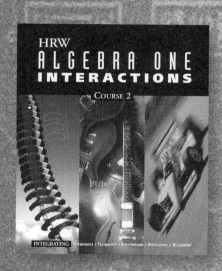

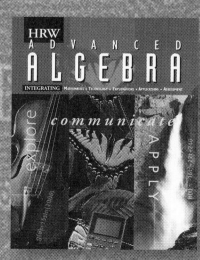

HOLT, RINEHART AND WINSTON
Harcourt Brace & Company

Austin • New York • Orlando • Atlanta • San Francisco • Boston • Dallas • Toronto • London

TABLE OF CONTENTS

Spreadsheets

f(g) Scholar™

Geometry Software

Hand-Held Computer

Appendix for Casio CFX-9850G

Index

Introduction to Technology in the Classroom

One popular dictionary defines *education* as the acquisition of both knowledge and skill. Implicit in this definition is the existence of an experience through which knowledge and skill are obtained and a facilitator who guides the process. That facilitator is you.

What ingredients does the educator incorporate into the educational experience? In particular, how does the mathematics teacher bring together methods and resources conducive to the development of the mathematical way of thinking?

For many years, the answer to these questions had been what some have called the lecture approach. According to this view, the teacher actively imparted knowledge to a passive listening audience via chalkboard demonstrations and explanations. Today's student, however, requires an active role in his or her development. Such a need partly explains the student's demand to know the relevance of mathematics and explains any lack of attention paid during lengthy explanations at the chalkboard.

In response to the changing profile of the mathematics student, educators have chosen to broaden the set of methods and techniques used in the mathematics classroom. The expanded list of strategies now includes the following: projects, small group cooperative learning, portfolios, alternative and performance assessment, student presentations, manipulative activities, explorations and investigations, and technology.

The focus and purpose of this book is as an introduction to various technologies used in today's mathematics classroom. Three basic types of technology are featured. They are:

- hand-held graphics calculators,

- computer software, and

- a hand-held computer, the TI-92.

The explanations that appear on the following pages are designed to show a teacher who is beginning to use technology both how the technology works and how the technology can be applied to productive mathematical ends.

In the case of graphics calculators, attention is paid to their use in the evaluation of various types of algebraic expressions, graphing various types of functions, operations on matrices, numerical and graphical analysis of different types of data, and calculation of combinatorics and probabilities. This book describes calculators made by Texas Instruments, Casio, Sharp and Hewlett Packard.

In the case of computer software, attention is paid to its use as a tool of exploration and discovery. This book describes:

- spreadsheet software (featuring *Microsoft® Excel* version 3.0),

- *f(g) Scholar™*,

- *The Geometer's Sketchpad®* (version 3.0),

- *Cabri Geometry II ™* (version 1.0), and

- *The Geometric superSupposer*.

In addition, this book provides a brief introduction to the TI-92, a hand-held computing device that integrates the capabilities of an advanced graphics calculator and computer software.

Introduction to Graphics Calculators in the Classroom

Perhaps you have heard of, seen, or even used an *abacus*. It is an ancient mechanical calculating device made of a frame containing sets of beads that slide along rods in the frame and is used to facilitate arithmetic calculations. The abacus was used in antiquity, the middle ages, and is still used today primarily in the Far East.

The early seventeenth century was the time when newer mechanical calculating devices were introduced. It was at this time that *circular and straight slide rules* gained acceptance and became widely used among mathematical practitioners. Markings on the rule were based on logarithmic scales. Versions made much later featured a sliding part of the rule that moved along a second or fixed part. With the slide rule, the user was able to make fairly rapid and complicated arithmetic calculations.

In the mid to latter part of this century, the modern *computer* was developed. At first, this electronic calculating device was so big that it filled a large room. As electronics technology advanced, new and powerful components made it possible to manufacture a smaller electronic calculating device, now known as the *pocket calculator*. This type of highly portable computing device ranges in complexity from simple (four-function calculators) to very sophisticated (advanced scientific and graphics calculators).

Throughout the history of mathematics, applied mathematics, science, and business, investigators and practitioners have done their work with whatever calculating devices the current technology had to offer. Since life in today's society has become so complicated and technical, there is a need to educate as many citizens as possible in the use of electronic devices relevant to virtually every career path. Perhaps this helps to explain the call for *algebra for everyone*.

It is no wonder that, given the rapid rate of change in society today, the availability of sophisticated hand-held calculators has caused a stir, perhaps even a revolution, in mathematical investigation, scientific application, commerce, technical trades, and yes, even in mathematics education itself.

What impact has the introduction of graphics calculators into the high school mathematics classroom had? Educators, administrators, and students are still struggling with an answer to that question. Some things do appear to be clear.

- The calculator lifts the burden of number crunching from the user.
- The graphics calculator makes it possible to join visualization with conceptualization.
- Some calculators make not only computations involving numbers possible, but also computations involving data, matrices, combinatorics and probability, and functions of many types.

There is, however, another impact. The technology coupled with societal need is exercising an influence on what topics will be part of the curriculum and on the amount of time these topics are taught.

Some greet the introduction of the graphics calculator with enthusiasm. Others express anxiety and uncertainty. What is explained and illustrated on the pages that follow will help those who are already interested and should help alleviate the anxiety of those who are just beginning.

Care and Resources for Graphics Calculators

Caring for Graphics Calculators

Welcome to the world of graphics calculators and their use in the mathematics classroom. In the course of your work with the activities that follow, you will, no doubt, notice that the graphics calculator can be a valuable resource in the teaching and learning process. Follow these simple steps to keep your calculators safe and in good working order.

Encourage students to:

- turn the calculator off when it is not in use.

- keep the protective plastic cover on it when it is not in use.

- keep pencils and sharp objects away from the liquid crystal display.

- wipe any spills on the keypad immediately by using a damp cloth.

- notify you about any low-battery signal they see on the display.

Because graphics calculators are expensive, you may also need to consider how you will mark them for identification and store them when they are not in use. For school-owned calculators, some teachers implement steps like the following.

- On the calculator back, mark each calculator with the school's name.

- Record the serial number and name of the student receiving each calculator.

- Collect calculators after use, and store them in a secure place.

Seeking Additional Calculator Resources

Calculator manufacturers have made available resources that include print material, technical support, and hardware add-ons.

Texas Instruments
  1-800-TI-CARES (842-2737)
  (one-on-one assistance)
  Texas Instruments
  PO Box 650311, M/S 3908
  Dallas, TX 75265
  e-mail: ti-cares@ti.com
  Newsletters: *TI-Cares* and *Eightysomething!*
Also available are:
- free color pictures and posters of calculators
- loan of calculators and overhead display
- workshops for a fee
- supplemental publications
- calculator-to-computer data link

Casio
  1-800-582-2763 (one-on-one assistance)
  Casio Education Division
  570 Mt. Pleasant Ave.
  Dover, NJ 07801
  e-mail: www.casio-usa.com
  Newsletter: *The Electronic Classroom*
Also available are:
- color pictures and free posters of calculators
- loan of calculators and overhead display
- free workshops
- teacher's resource books
- calculator-to-computer data link

Sharp
  1-201-529-6340 (inservice workshops)
  1-201-529-9542 (loan program and posters)
  Sharp Educational Division
  Box G
  Sharp Plaza
  Mahwah, NJ 07430-2135
Also available are:
- free posters
- loan of calculators and overhead display
- inservice workshops

Hewlett Packard
  1-800-795-7177 (education orders)
  HP Education Programs
  1000 N.E. Circle Blvd.
  Corvallis, OR 97330-4239
  Newsletter: *HP 38G ClassNotes*
Also available are:
- free color pictures and posters of calculators
- loan of calculators and overhead display
- inservice workshops
- calculator-to-computer data link

GETTING ACQUAINTED

Familiarizing Yourself with the Calculator

The graphics calculator you will be using is a technological device that operates on battery power. As such, you will need to check out various aspects of your calculator as an electronic machine. In the descriptions that follow, you will see how to turn the calculator on, turn it off, reset its memory, and change the contrast on the display.

	ON/OFF	RESET	TO CHANGE CONTRAST ON THE DISPLAY
Texas Instruments **TI-83**	ON [ON] OFF [2nd] [ON]	[2nd] [+ MEM] **5: Reset...** [ENTER] **2: Defaults...** [ENTER]	[2nd] [▲] more [▼] less
Texas Instruments **TI-82**	ON [ON] OFF [2nd] [ON]	[2nd] [+ MEM] **3: Reset...** [ENTER] **2: Reset**	[2nd] [▲] more [▼] less
Texas Instruments **TI-81**	ON [ON] OFF [2nd] [ON]	[2nd] [+ RESET] **2: Reset...** [ENTER]	[2nd] [▲] more [▼] less
Casio **CFX-9850G**	ON [AC/ON] OFF [SHIFT] [AC/ON]	[MENU] **E MEM** [EXE] **RESET** [EXE]	[MENU] **D CONT** [EXE] **COLOR CONTRAST** [EXE] [◄] or [►] [EXE]
Casio **CFX-9800G**	ON [AC/ON] OFF [SHIFT] [AC/ON]	[MENU] **C OPTION** [EXE] **RESET** [EXE]	[MENU] **C OPTION** [EXE] **COLOR CONTRAST** [EXE] [◄] or [►] [EXE]
Casio **fx-9700GE**	ON [AC/ON] OFF [SHIFT] [AC/ON]	[MENU] **D RESET** [EXE] [F1]	[MENU] **C CONT** [EXE] [◄] LIGHT [►] DARK
Casio **fx-7700GE**	ON [AC/ON] OFF [SHIFT] [AC/ON]	[MENU] **B RESET** [EXE] [F1]	[MENU] **A CONT** [EXE] [◄] LIGHT [►] DARK
Sharp **EL-9200C** **EL-9300C**	ON [ON] OFF [2nd F] [ON]	Press the RESET button on the back of the calculator.	[2nd F] [OPTION MENU] **A CTRST** [+] DARK [−] LIGHT
Hewlett Packard **HP 38G**	ON [ON] OFF [▨] [ON]	RESET [SETUP PLOT] [▨] [CLEAR DEL]	Simultaneously press [ON] and [+] more or [−] less

A View of the Texas Instruments TI-83

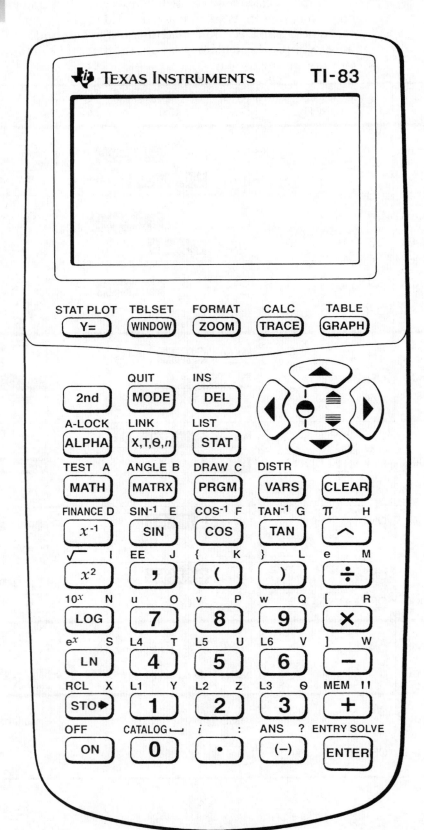

Choosing and Accessing TI-83 Menus

On the TI-83, you will find a variety of menus to choose from. The selection of a particular menu depends on the task you wish to perform.

You can access each menu that is described on this page by following each key sequence given. Each diagram shows the display that results from pressing the indicated keys. Once you are in a menu that contains options, use ◄ , ▲ , ► , and ▼ to move from one option to another. Press ENTER when you have finished making your selections.

To make calculations, evaluate formulas, and evaluate expressions:

2nd [QUIT] MODE

To choose a function type, angle measurement system, graph style, or complex numbers:

2nd [QUIT] MODE MODE

To determine the portion of the coordinate plane in which your graph(s) will be displayed and set the *x*- and *y*-scales:

2nd [QUIT] MODE WINDOW

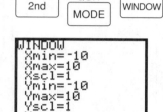

To enter, or define, one or more functions to be graphed; also, to turn stat plots on or off and set graph style:

2nd [QUIT] MODE Y=

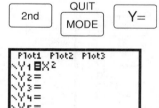

To enter one or more matrices, edit a matrix, delete a matrix, and perform operations on matrices:

2nd [QUIT] MODE MATRX

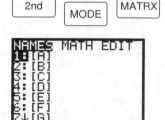

To enter one-variable data, two-variable data, edit data, calculate statistical measures, or select inferential statistics:

2nd [QUIT] MODE STAT

The actual settings you see on a particular graphics calculator display may vary from those shown here, since different settings may have been made and saved on a previous occasion.

On the pages that follow, you will see how these menus are used to perform arithmetic and algebraic computations, graph both explicit and parametric functions, solve systems of equations, find zeros of functions, work with matrices, and work with data.

A View of the Texas Instruments TI-82

TEXAS INSTRUMENTS *TI-82*

STAT PLOT	TblSet		CALC	TABLE
Y=	WINDOW	ZOOM	TRACE	GRAPH

	QUIT	INS		
2nd	MODE	DEL	▲	►
A-LOCK	LINK	LIST	◄	▼
ALPHA	X,T,Θ	STAT		

TEST A	ANGLE B	DRAW C	Y-VARS	
MATH	MATRX	PRGM	VARS	CLEAR

ABS D	SIN⁻¹ E	COS⁻¹ F	TAN⁻¹ G	π H
x^{-1}	SIN	COS	TAN	^

√ I	EE J	{ K	} L	M
x^2	,	(	)	÷

10^x N	Un-1 O	Vn-1 P	n Q	[R
LOG	7	8	9	×

e^x S	L4 T	L5 U	L6 V	] W
LN	4	5	6	−

RCL X	L1 Y	L2 Z	L3 Θ	MEM ¶¶
STO►	1	2	3	+

OFF	⌐	:	ANS ?	ENTRY
ON	0	•	(−)	ENTER

Choosing and Accessing TI-82 Menus

On the TI-82, you will find a variety of menus to choose from. The selection of a particular menu depends on the task you wish to perform.

You can access each menu that is described on this page by following each key sequence given. Each diagram shows the display that results from pressing the indicated keys. Once you are in a menu that contains options, use ◄ , ▲ , ► , and ▼ to move from one option to another. Press ENTER when you have finished making your selections.

To make calculations, evaluate formulas, and evaluate expressions:

2nd	QUIT MODE

■

To choose a function type, angle measurement system, and graph style:

2nd	QUIT MODE	MODE

```
Normal Sci Eng
Float 0123456789
Radian Degree
Func Par Pol Seq
Connected Dot
Sequential Simul
FullScreen Split
```

To determine the portion of the coordinate plane in which your graph(s) will be displayed and set the *x*- and *y*-scales:

2nd	QUIT MODE	WINDOW

```
WINDOW FORMAT
Xmin=-4.7
Xmax=4.7
Xscl=1
Ymin=-3.1
Ymax=3.1
Yscl=1
```

To enter, or define, one or more functions to be graphed:

2nd	QUIT MODE	Y=

```
Y1=■
Y2=
Y3=
Y4=
Y5=
Y6=
Y7=
Y8=
```

To enter one or more matrices, edit a matrix, delete a matrix, and perform operations on matrices:

2nd	QUIT MODE	MATRX

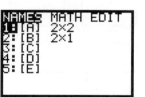
```
NAMES MATH EDIT
1:[A]  2×2
2:[B]  2×1
3:[C]
4:[D]
5:[E]
```

To enter one-variable data, two-variable data, edit data, calculate statistical measures, and display statistical graphs:

2nd	QUIT MODE	STAT

```
EDIT CALC
1:Edit…
2:SortA(
3:SortD(
4:ClrList
```

The actual settings you see on a particular graphics calculator display may vary from those shown here as different settings may have been made and saved on a previous occasion.

On the pages that follow, you will see how these menus are used to perform arithmetic and algebraic computations, graph both explicit and parametric functions, solve systems of equations, find zeros of functions, work with matrices, work with data, and so on.

A View of the Texas Instruments TI-81

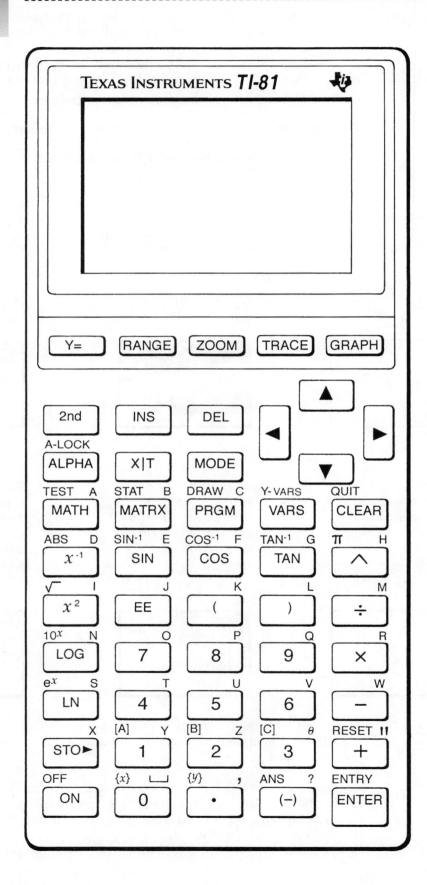

Choosing and Accessing TI-81 Menus

On the TI-81, you will find a variety of menus to choose from. The selection of a particular menu depends on the task you wish to perform.

You can perform each task that is described on this page by following each key sequence given. Each diagram shows the display that results from pressing the indicated keys. Once you are in a menu that contains options, use $\boxed{\blacktriangleleft}$, $\boxed{\blacktriangle}$, $\boxed{\blacktriangleright}$, and $\boxed{\blacktriangledown}$ to move from one option to another. Press $\boxed{\text{ENTER}}$ when you have finished making your selections.

To make calculations, evaluate formulas, and evaluate expressions:

$\boxed{\text{2nd}}$ $\boxed{\substack{\text{QUIT} \\ \text{CLEAR}}}$

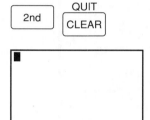

To choose a function type, angle measurement system, and graph style:

$\boxed{\text{2nd}}$ $\boxed{\substack{\text{QUIT} \\ \text{CLEAR}}}$ $\boxed{\text{MODE}}$

To determine the portion of the coordinate plane in which your graph(s) will be displayed and set the *x*- and *y*-scales:

$\boxed{\text{2nd}}$ $\boxed{\substack{\text{QUIT} \\ \text{CLEAR}}}$ $\boxed{\text{RANGE}}$

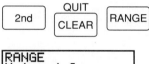

To enter, or define, one or more functions to be graphed:

$\boxed{\text{2nd}}$ $\boxed{\substack{\text{QUIT} \\ \text{CLEAR}}}$ $\boxed{\text{Y=}}$

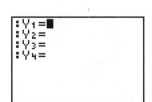

To enter one or more matrices, edit a matrix, delete a matrix, and perform operations on matrices:

$\boxed{\text{2nd}}$ $\boxed{\substack{\text{QUIT} \\ \text{CLEAR}}}$ $\boxed{\text{MATRX}}$

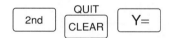

To enter one-variable data, two-variable data, edit data, and calculate statistical measures:

$\boxed{\text{2nd}}$ $\boxed{\substack{\text{QUIT} \\ \text{CLEAR}}}$ $\boxed{\text{2nd}}$

$\boxed{\substack{\text{STAT B} \\ \text{MATRX}}}$

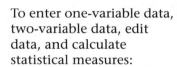
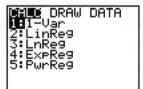

The actual settings you see on a particular graphics calculator display may vary from those shown here as different settings may have been made and saved on a previous occasion.

On the pages that follow, you will see how these menus are used to perform arithmetic and algebraic computations, graph both explicit and parametric functions, solve systems of equations, find zeros of functions, work with matrices, and work with data.

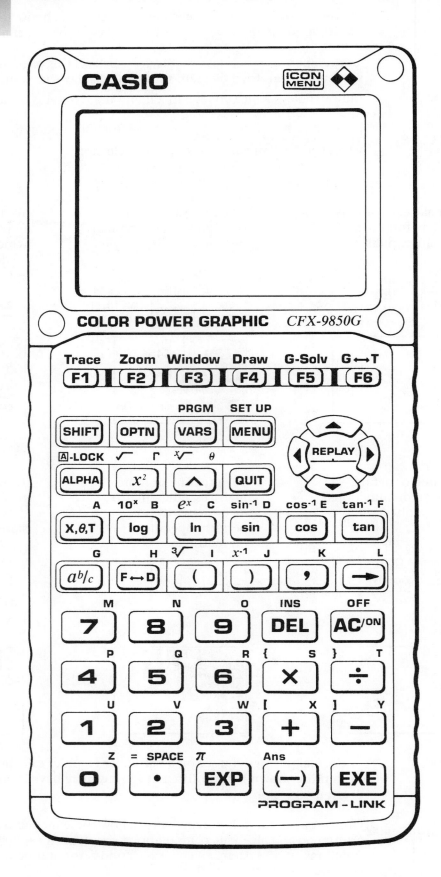

CASIO CFX-9850G

Choosing and Accessing CFX-9850G Menus

On the Casio CFX-9850G, you will find a variety of menus to choose from. The selection of a particular menu depends on the task you wish to perform. You can perform each task that is described on this page by following each key sequence given. Each diagram shows the display that results. Once in a menu that contains options, use ◄, ▲, ►, and ▼ to move from one option to another.

To begin:

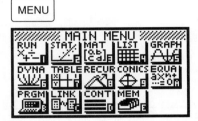

To choose a function type and angle measurement system:

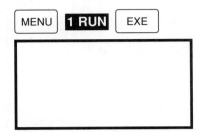

To enter, or define, one or more functions to be graphed:

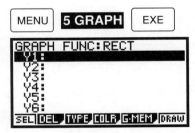

To enter matrices and operate on them

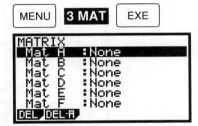

To solve equations and linear systems:

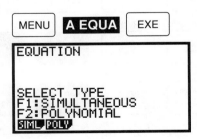

To enter data, edit it, calculate statistical measures, and display statistical graphs:

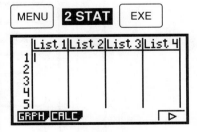

The actual settings you see on a particular graphics calculator display may vary from those shown here as different settings may have been made and saved on a previous occasion.

A View of the Casio CFX-9800G

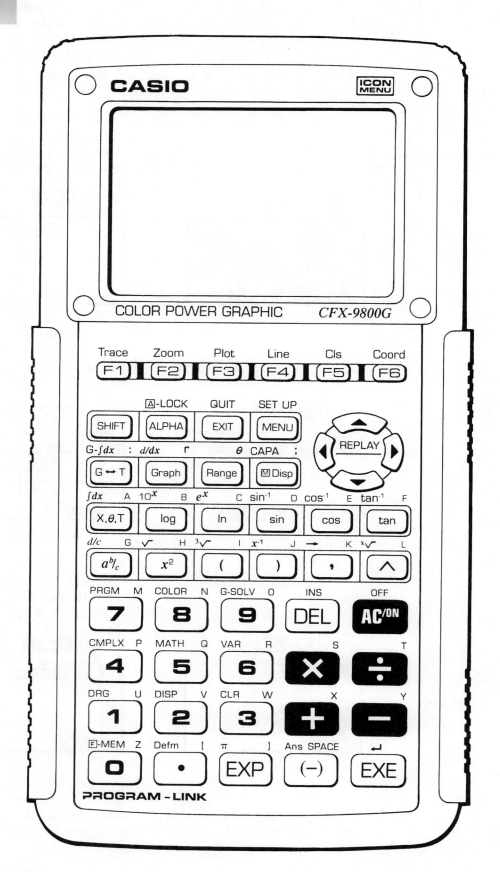

CASIO CFX-9800G

Choosing and Accessing CFX-9800G Menus

On the Casio CFX-9800G, you will find a variety of menus to choose from. The selection of a particular menu depends on the task you wish to perform. You can perform each task that is described on this page by following each key sequence given. Each diagram shows the display that results. Once in a menu that contains options, use ◄ , ▲ , ► , and ▼ to move from one option to another.

To begin:

MENU

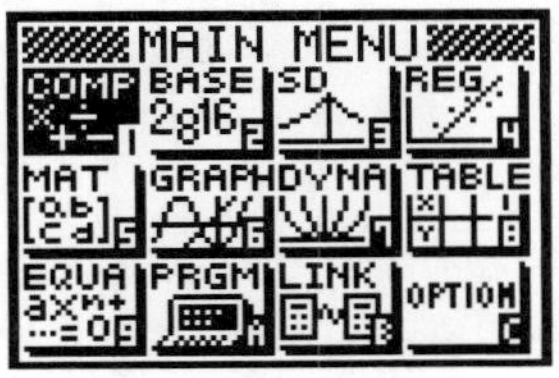

To choose a function type and angle measurement system:

MENU **1 COMP** EXE

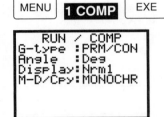

To enter, or define, one or more functions to be graphed:

MENU **6 GRAPH** EXE

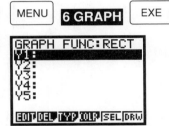

To enter matrices and operate on them:

MENU **5 MAT** EXE

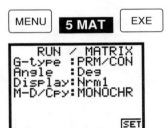

To solve equations and linear systems:

MENU **9 EQUA** EXE

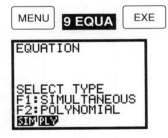

To enter data (one-variable (**SD**) or two-variable (**REG**)), edit it, calculate statistical measures, and display statistical graphs:

MENU **3 SD** EXE

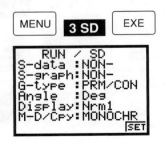

MENU **4 REG** EXE

The actual settings you see on a particular graphics calculator display may vary from those shown here as different settings may have been made and saved on a previous occasion.

A View of the Casio fx-9700GE

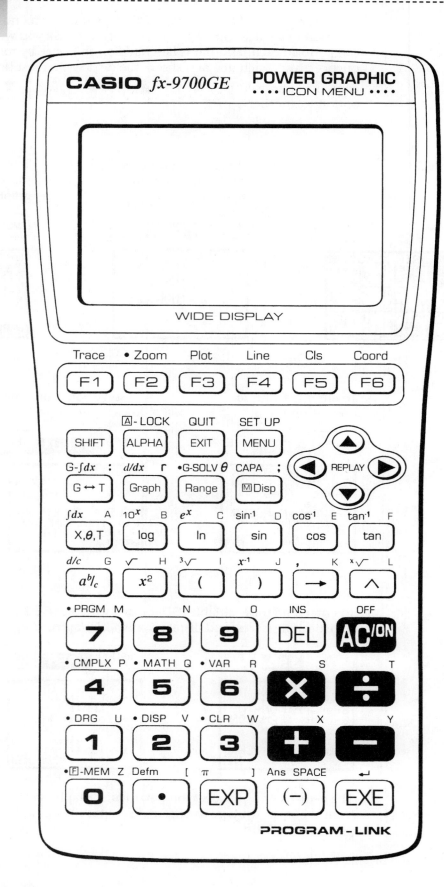

Choosing and Accessing fx-9700GE Menus

On the Casio fx-9700GE, you will find a variety of menus to choose from. The selection of a particular menu depends on the task you wish to perform. You can perform each task that is described on this page by following each key sequence given. Each diagram shows the display that results. Once in a menu that contains options, use $\boxed{\blacktriangleleft}$, $\boxed{\blacktriangle}$, $\boxed{\blacktriangleright}$, and $\boxed{\blacktriangledown}$ to move from one option to another.

To begin:

$\boxed{\text{MENU}}$

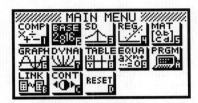

To choose a function type and angle measurement system:

$\boxed{\text{MENU}}$ **1 COMP** $\boxed{\text{EXE}}$

To enter, or define, one or more functions to be graphed:

$\boxed{\text{MENU}}$ **6 GRAPH** $\boxed{\text{EXE}}$

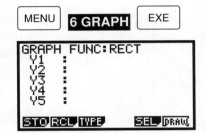

To enter matrices and operate on them:

$\boxed{\text{MENU}}$ **5 MAT** $\boxed{\text{EXE}}$

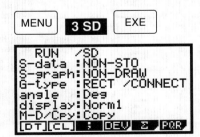

To solve equations and linear systems:

$\boxed{\text{MENU}}$ **9 EQUA** $\boxed{\text{EXE}}$

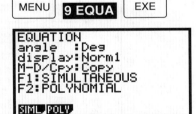

To enter data (one-variable (**SD**) or two-variable (**REG**)), edit it, calculate statistical measures, and display statistical graphs:

$\boxed{\text{MENU}}$ **3 SD** $\boxed{\text{EXE}}$

$\boxed{\text{MENU}}$ **4 REG** $\boxed{\text{EXE}}$

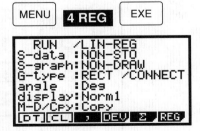

The actual settings you see on a particular graphics calculator display may vary from those shown here as different settings may have been made and saved on a previous occasion.

A View of the Casio fx-7700GE

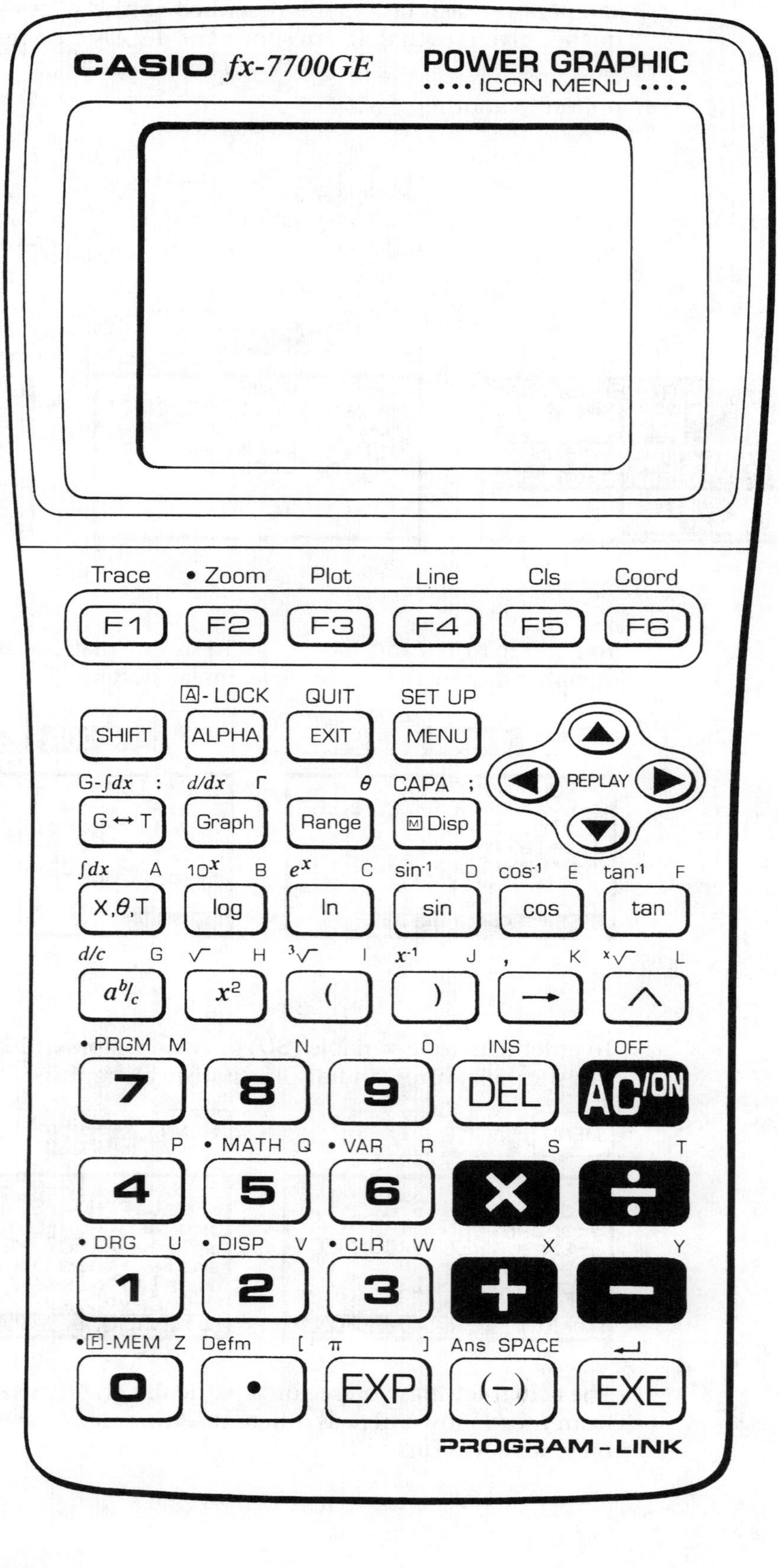

CASIO *fx-7700GE*   **POWER GRAPHIC**
···· ICON MENU ····

Trace	• Zoom	Plot	Line	Cls	Coord
F1	F2	F3	F4	F5	F6

Ⓐ- LOCK   QUIT   SET UP

SHIFT   ALPHA   EXIT   MENU   ▲ ◀ REPLAY ▶ ▼

G-∫dx :   d/dx Γ   θ   CAPA ;

G↔T   Graph   Range   Ⓜ Disp

∫dx A 10^x B e^x C $\sin^{-1}$ D $\cos^{-1}$ E $\tan^{-1}$ F

X,θ,T   log   ln   sin   cos   tan

d/c G   √ H $^3\sqrt{}$ I x^{-1} J   , K $^x\sqrt{}$ L

$a^b/_c$ x^2 (   )   →   ∧

• PRGM M   N   O   INS   OFF

7   8   9   DEL   AC$^{/ON}$

P   • MATH Q   • VAR R   S   T

4   5   6   ×   ÷

• DRG U   • DISP V   • CLR W   X   Y

1   2   3   +   −

• Ⓕ-MEM Z   Defm   [π]   Ans SPACE   ↵

0   •   EXP   (−)   EXE

PROGRAM – LINK

Choosing and Accessing fx-7700GE Menus

On the Casio fx-7700GE, you will find a variety of menus to choose from. The selection of a particular menu depends on the task you wish to perform. You can perform each task that is described on this page by following each key sequence given. Each diagram shows the display that results. Once in a menu that contains options, use ◄ , ▲ , ► , and ▼ to move from one option to another.

To begin:

MENU

To choose a function type and angle measurement system:

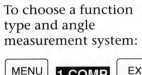

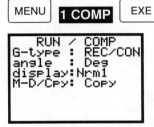

To enter, or define, one or more functions to be graphed:

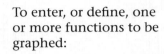

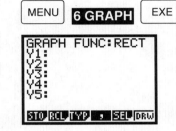

To enter matrices and operate on them:

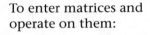

To solve equations and linear systems:

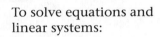

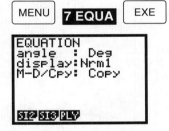

To enter data (one-variable (**SD**) or two-variable (**REG**)), edit it, calculate statistical measures, and display statistical graphs:

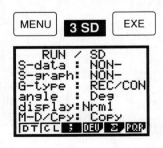

The actual settings you see on a particular graphics calculator display may vary from those shown here as different settings may have been made and saved on a previous occasion.

A View of the Sharp EL-9300C and EL-9200C

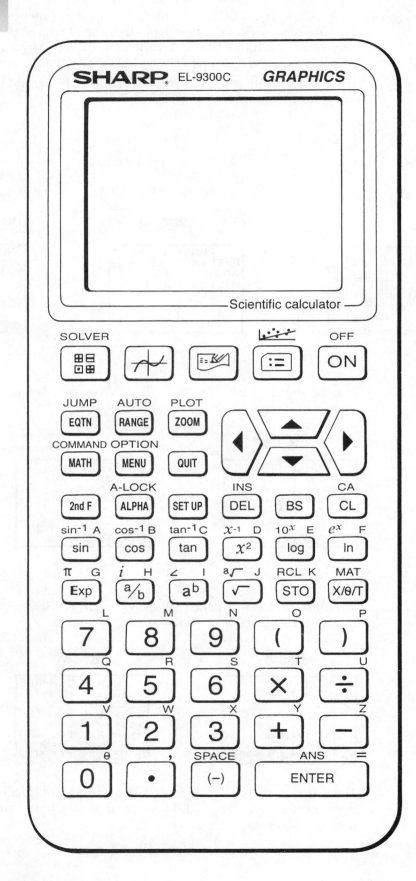

Choosing and Accessing EL-9300C and EL-9200C Menus

On the Sharp graphics calculators, you will find a variety of menus to choose from. The selection of a particular menu depends on the task you wish to perform.

You can access each menu that is described on this page by following each key sequence given. Each diagram shows the display that results from pressing the indicated keys. Once in a menu that contains options, use ◄ , ▲ , ► , and ▼ to move from one option to another. Press ENTER when you have finished making your selections.

To make calculations, evaluate formulas, and evaluate expressions:

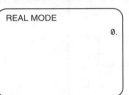

To choose an angle measurement system and a coordinate system for graphing functions:

SET UP

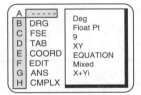

To determine the portion of the coordinate plane in which your graph(s) will be displayed and to set the scales for the *x*- and *y*-axes:

 Range

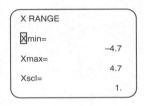

To find coordinates of important points on the graph of a function when a graph is displayed:

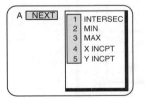

To enter one-variable or two variable data when the calculator does not currently contain data:

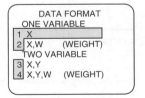

To enter one-variable or two variable data when the calculator already contains data:

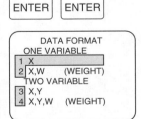

The actual settings you see on a particular graphics calculator display may vary from those shown here as different settings may have been made and saved on a previous occasion.

On the pages that follow, you will see how these menus are used to perform arithmetic and algebraic computations, graph both explicit and parametric functions, solve systems of equations, find zeros of functions, work with matrices, work with data, and so on.

A View of the Hewlett Packard HP 38G

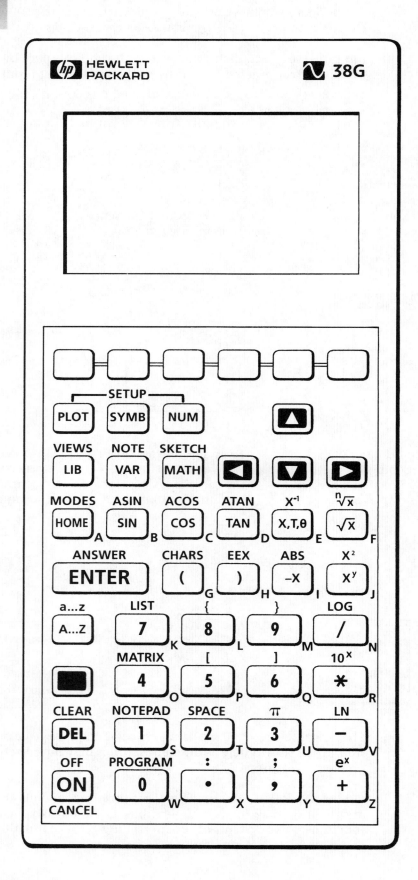

Choosing and Accessing HP 38G Menus

On the HP 38G, you will find a variety of menus to choose from. The selection of a particular menu depends on the task you wish to perform.

You can access each menu that is described on this page by following each key sequence given. Each diagram shows the display that results from pressing the indicated keys. Once in a menu that contains options, use ◀ , ▲ , ▶ , and ▼ to move from one option to another. Press ENTER when you have finished making your selections.

To make calculations, evaluate formulas, and evaluate expressions:

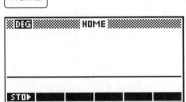

To choose a function type, a solver, and statistical options:

To access mathematical functions, statistical functions, symbolic options, and tests:

To determine that portion of the coordinate plane in which your graph(s) will be displayed and set the *x*- and *y*-scales:

To choose an angle measurement system:

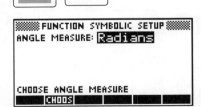

To zoom in on a portion of the coordinate plane that contains important characteristics of one or more graphs:

The actual settings you see on a particular graphics calculator display may vary from those shown here as different settings may have been made and saved on a previous occasion.

On the pages that follow, you will see how these menus are used to perform arithmetic and algebraic computations, graph both explicit and parametric functions, solve systems of equations, find zeros of functions, work with matrices, work with data, and so on.

GRAPHICS CALCULATOR SKILLS

Evaluating Expressions

A graphics calculator is a powerful tool that enables you to visualize functions and data. It is, of course, also a scientific calculator. As such, you can use it to handle algebraic and numerical tasks, such as the evaluation of an expression.

An *algebraic expression* is an expression that involves variables and numbers joined or related by mathematical operations. When numerical values are substituted for the variables, the resulting expression is called a *numerical expression*.

Suppose that you want to find the volume of a sphere whose radius is 7 units. You will need the fact that a sphere with radius r units has volume $V = \frac{4}{3}\pi r^3$ cubic units and will then need to evaluate the numerical expression $\frac{4}{3}\pi(7^3)$.

Skill 1: Evaluating a Formula From Geometry

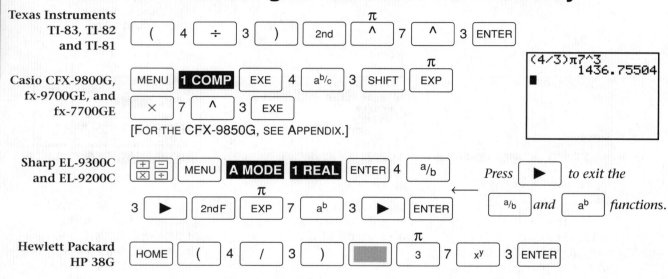

To evaluate the volume formula for $r = 8.5$, you need only edit the formula already entered.

Skill 2: Editing and Reevaluating a Formula

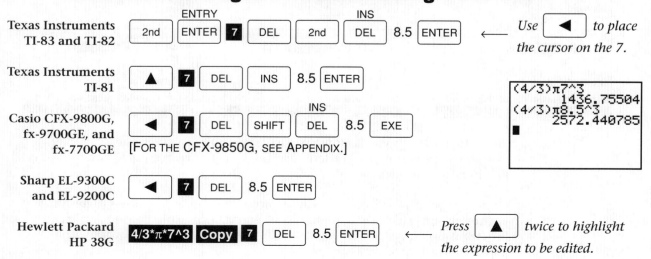

One of the most famous formulas in algebra is the *quadratic formula*. It is the following. If $ax^2 + bx + c = 0$, then $x = \dfrac{-b \pm \sqrt{b^2 - 4ac}}{2a}$. The following key sequences show the use of this formula applied to $2x^2 - 5x - 3 = 0$.

- Since the formula involves $\pm$, you will need to evaluate the formula using $+$, and then edit it to use $-$.

$$\frac{-(-5) + \sqrt{(-5)^2 - 4(2)(-3)}}{2(2)}$$

- Since the formula involves a complicated quotient, you will need to enclose both numerator and denominator in parentheses.

$$\left(-(-5) + \sqrt{(-5)^2 - 4 \times 2 \times (-3)}\right) \div (2 \times 2)$$

$$\left(--5 + \sqrt{\left((-5)^2 - 4 \times 2 \times (-3)\right)}\right) \div (2 \times 2)$$

- Since the formula involves a square root and the radicand is complicated, the radicand must also be enclosed in parentheses.

Skill 3: Evaluating a Formula from Algebra

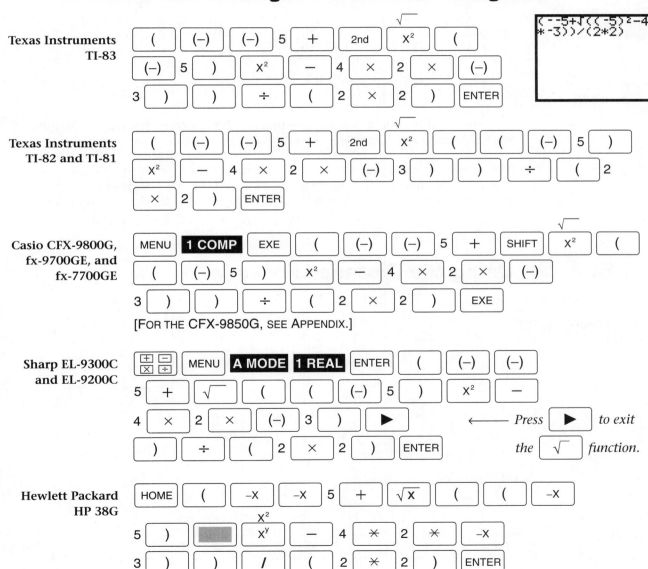

[FOR THE CFX-9850G, SEE APPENDIX.]

GRAPHICS CALCULATOR SKILLS

Investment problems involving compound interest occur in mathematics courses and in everyday life. The expression $1250(1.04)^{1\frac{3}{4}}$ indicates the amount to which \$1250 will grow over a period of $1\frac{3}{4}$ years if the annual interest rate is 4%, the interest is compounded annually, and no money is withdrawn from the account. The keystrokes that follow illustrate the evaluation of this expression.

Skill 4: Evaluating an Exponential Expression

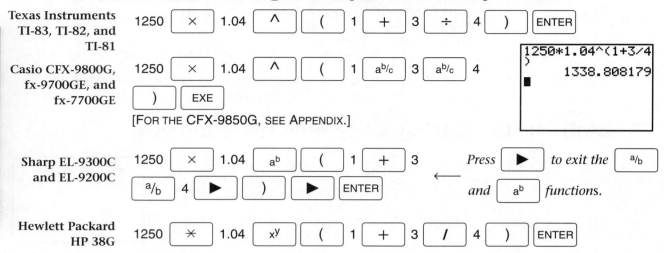

Texas Instruments TI-83, TI-82, and TI-81

Casio CFX-9800G, fx-9700GE, and fx-7700GE

[FOR THE CFX-9850G, SEE APPENDIX.]

Sharp EL-9300C and EL-9200C

Hewlett Packard HP 38G

For more information about expressions involving exponents and logarithms, see *Exponential and Logarithmic Functions* (pages 56-58).

The *distance formula* provides an algebraic means for finding the distance between two points whose coordinates are given. If points P and Q have coordinates $P(a, b)$ and $Q(c, d)$, then the distance PQ between them is given by the formula $PQ = \sqrt{(c-a)^2 + (d-b)^2}$. Suppose you have $P(-2, -1)$ and $Q(6, 4)$. Skill 5 shows how to find PQ.

Skill 5: Using the Distance Formula

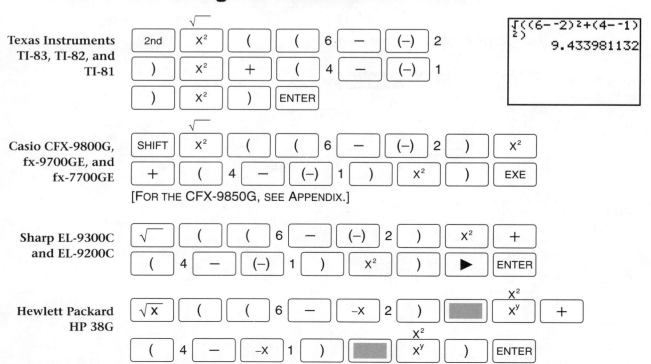

Texas Instruments TI-83, TI-82, and TI-81

Casio CFX-9800G, fx-9700GE, and fx-7700GE

[FOR THE CFX-9850G, SEE APPENDIX.]

Sharp EL-9300C and EL-9200C

Hewlett Packard HP 38G

The equation $x = \dfrac{20 \sin 32.6°}{\sin 47.2°}$ can be used to find x in the triangle shown here. Carry out the appropriate key sequence that follows to approximate x. Since the expression involves angle measures in degrees, select degree mode before beginning the evaluation.

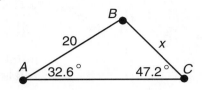

Skill 6: Evaluating a Trigonometric Expression

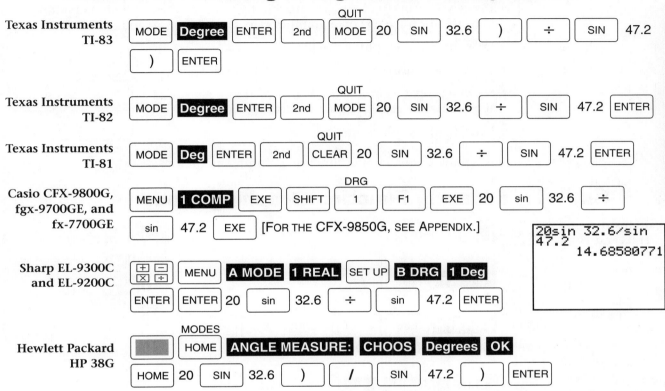

See *Trigonometric Functions* (pages 62-65) for more information.

INVESTIGATIONS

1. Consider $f(x) = \dfrac{x+1}{x+1}$. Suppose $x = 1$.

   a. Find the value of the expression after the substitution for x is made.

   b. What happens if you evaluate the expression without using parentheses to group the numerator? the denominator? either the numerator or denominator?

   c. Try exploring other expressions that have one set of values if parentheses are used and a different set of values if parentheses are not used.

2. Suppose you want to solve $2.5x + 3.2 = 1.5c$ for x given various different but fixed values of c.

   a. An expression for x in terms of c can be called a *calculator-ready solution* for x. Write x in terms of c. Find x given $c = 1$ and $c = 2$.

   $$\left[x = \frac{1.5c - 3.2}{2.5}; \text{ If } c = 1, x = -0.68, \text{ If } c = 2, x = -0.08 \right]$$

   b. Try investigating other equations involving x and c that you can solve in a similar fashion.

GRAPHICS CALCULATOR SKILLS

The Graph Viewing Window

The *graph* of a function $y = f(x)$ is the set of all points $P(x, y)$ in the coordinate plane that satisfy the equation. If the domain of the function is the set of all real numbers, then the graph of the function will be infinite in extent. No picture will be able to show the entire graph. A *representative, characteristic*, or *complete graph* is a portion of the entire graph that shows, in a finite rectangle, all of the important features of the graph.

Suppose you want to obtain a representative or characteristic graph of a function of the form $y = f(x)$ on a graphics calculator. There are four steps involved.

- Enter function graphing mode.

- Type the expression that defines the function.

- Select a proper *viewing window*, in which a representative graph of the function appears.

- Execute the graph command.

Skill 1 shows how to enter function graphing mode.

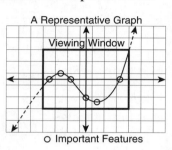

A Representative Graph
Viewing Window
O Important Features

Skill 1: Entering Function Graphing Mode

Texas Instruments TI-83 and TI-82
MODE | Func | ENTER | Y=

Texas Instruments TI-81
MODE | Function | ENTER | Y=

Casio CFX-9800G, fx-9700GE, and fx-7700GE
MENU | 6 GRAPH | EXE | SHIFT | SET UP MENU | F1 | EXIT
[FOR THE CFX-9850G, SEE APPENDIX.]

Sharp EL-9300C and EL-9200C
↵ | SET UP | E COORD | 1 XY | ENTER | ENTER | EQTN

Hewlett Packard HP 38G
LIB | Function | ENTER

```
Norm Sci Eng
Float 0123456789
Rad Deg
Function Param
Connected Dot
Sequence Simul
Grid Off Grid On
Rect Polar
```

In some cases a symmetric viewing window will do nicely. A *symmetric viewing window* is one in which the origin of the coordinate system is at the center of the display. At other times, you will need to use an *asymmetric viewing window*, one in which the origin of the system is somewhere other than at the center of the display. These calculator displays illustrate the distinction between a symmetric and an asymmetric viewing window.

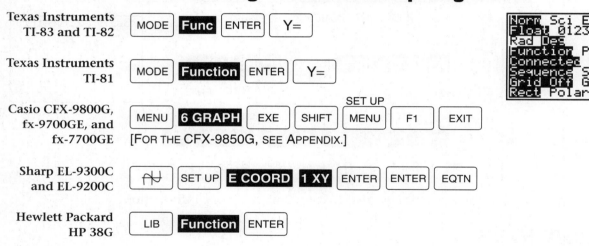

$$f(x) = 0.25x^3 - 2.5x$$

$$f(x) = (x - 4)^2 - 4$$

Three options for the selection of a viewing window are available.

- Accept the current viewing window.
- Define your own viewing window based on limits for the variables *x* and *y*.
- Select a preset, or built-in, viewing window.

Before making a definite choice of viewing window, you may wish to examine the settings for the current viewing window.

Skill 2: Displaying the Current Window Settings

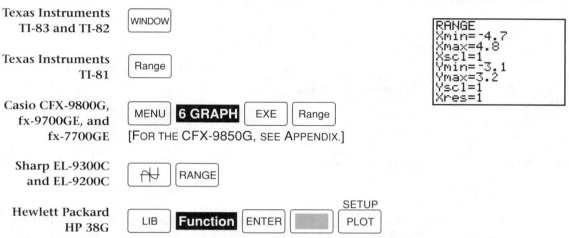

If the current viewing window settings are acceptable, you are ready to graph your function. If you want to define your own viewing window to be, for example, bounded by $-6 \le x \le 6$ and $-4 \le y \le 4$ with each axis having a scale of 1, you will need to follow the appropriate key sequence outlined here.

Skill 3: Defining Your Viewing Window

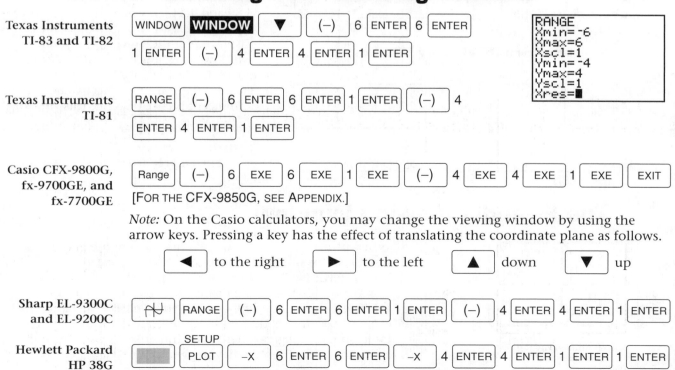

There are times when you might wish to use one of the preset or built-in viewing windows. Use the down arrow to highlight the indicated menu choice, such as � **4: ZDecimal**, and press ENTER to select it.

Skill 4: Choosing a Preset Window

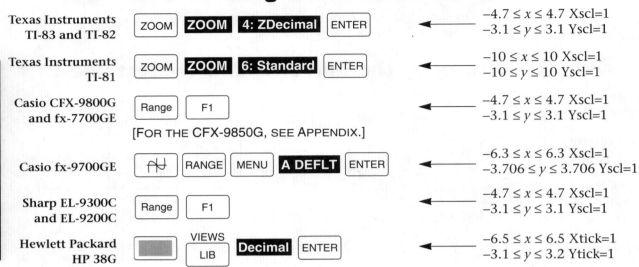

Texas Instruments TI-83 and TI-82	ZOOM **ZOOM** **4: ZDecimal** ENTER ◄——	$-4.7 \le x \le 4.7$ Xscl=1 $-3.1 \le y \le 3.1$ Yscl=1
Texas Instruments TI-81	ZOOM **ZOOM** **6: Standard** ENTER ◄——	$-10 \le x \le 10$ Xscl=1 $-10 \le y \le 10$ Yscl=1
Casio CFX-9800G and fx-7700GE	Range F1 ◄—— [FOR THE CFX-9850G, SEE APPENDIX.]	$-4.7 \le x \le 4.7$ Xscl=1 $-3.1 \le y \le 3.1$ Yscl=1
Casio fx-9700GE	⚏ RANGE MENU **A DEFLT** ENTER ◄——	$-6.3 \le x \le 6.3$ Xscl=1 $-3.706 \le y \le 3.706$ Yscl=1
Sharp EL-9300C and EL-9200C	Range F1 ◄——	$-4.7 \le x \le 4.7$ Xscl=1 $-3.1 \le y \le 3.1$ Yscl=1
Hewlett Packard HP 38G	▨ VIEWS LIB **Decimal** ENTER ◄——	$-6.5 \le x \le 6.5$ Xtick=1 $-3.1 \le y \le 3.2$ Ytick=1

If the current viewing window settings are acceptable, you are ready to graph your function. For instance, you may wish to graph the function $y = 0.5x - 2$ and find the exact value of its x-intercept. The screens displayed for the TI-81 Standard viewing window show two consecutive cursor traces. (For more information on tracing, see *The Tracing and Zooming Operations*, pages 30–32.) Notice, from the bottom of the calculator displays shown, that the x-intercept is between 3.89 and 4.11. However, the object is to find the *exact* x-intercept. The screens displayed for the TI-81 friendly viewing window also show consecutive cursor traces. In this case, each x-value increases by 0.1 and the exact value of the x-intercept is (4.0, 0). Since this viewing window allows TRACE to cursor by even increments of x, 3.9, 4.0, 4.1, …, it is called a **friendly** viewing window.

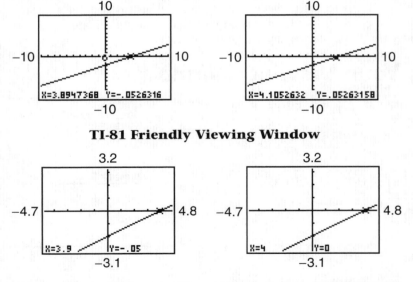

TI-81 Standard Viewing Window

TI-81 Friendly Viewing Window

The Friendly Viewing Window

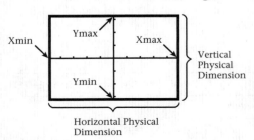

Xmin, Ymax, Xmax, Ymin

Horizontal Physical Dimension

Vertical Physical Dimension

When using the trace feature to cursor along the graph of a function, it is often helpful to use a friendly viewing window. These viewing window settings depend on the actual dimensions or usable size of the display on your particular calculator. These dimensions are designated by *pixels*, the basic units of picture elements that make up the calculator display. The table on page 29 gives these pixel dimensions with the corresponding *preset* friendly viewing windows.

Calculator	Physical Viewing Window Dimensions (Horiz. by Vert.)	Preset Friendly Viewing Window
TI-82, TI-83 Casio CFX-9800G, fx-7700GE Sharp EL-9300C, EL-9200C	94 pixels by 62 pixels	Xmin: –4.7, Xmax: 4.7 Ymin: –3.1, Ymax: 3.1
TI-81 * The TI-81 does *not* have a *preset* friendly viewing window.	95 pixels by 63 pixels	Xmin: –4.7, Xmax: 4.8 * Ymin: –3.1, Ymax: 3.2
Casio CFX-9850G	126 pixels by 62 pixels	Xmin: –6.3, Xmax: 6.3 Ymin: –3.1, Ymax: 3.1
Casio fx-9700GE	126 pixels by 74.12 pixels	Xmin: –6.3, Xmax: 6.3 Ymin: –3.706, Ymax: 3.706
HP 38G	130 pixels by 63 pixels	Xmin: –6.5, Xmax: 6.5 Ymin: –3.1, Ymax: 3.2

Although the preset friendly viewing windows are convenient, an infinite number of friendly viewing windows are possible. The only requirement is that the horizontal window settings correspond to the horizontal pixel dimension of your calculator display. For example, you may want to find a friendly viewing window for the cosine function over the domain –360° to 360° and over the range –1 to 1. Multiply the Xmin and Xmax from the table above by a whole number so that the required domain would be included. Notice that 100 times Xmin and Xmax will satisfy this requirement. These settings will allow the trace feature to cursor by exact degree values, in increments of 10°. On the TI-81, this window would be: Xmin = –470, Xmax = 480, Xscl = 90, Ymin = –1.1, Ymax = 1.1, and Yscl = 0.1. Notice that Xmax – Xmin = 480 – (–470) = 950 = 10 · 95, or 10 times the horizontal pixel dimension. A friendly viewing window on any calculator will have this characteristic, that Xmax – Xmin = k · the horizontal pixel dimension, where k is a positive rational number.

INVESTIGATIONS

1. a. Graph the function $f(x) = x^2 – 1$ using the friendly viewing window in the table above. Trace to find its vertex and its x- and y-intercepts. This viewing window shows a representative graph of the function since the vertex and both intercepts are visible. [The vertex is (0, 1), the x-intercepts are (–1, 0) and (1, 0), and the y-intercept is (0, –1).]

b. Graph the function $g(x) = (x – 2)^2 – 9$ using the same viewing window. This viewing window is *not* representative since the vertex and one x-intercept are not visible. Adjust the window settings to display a friendly representative graph, and trace to find the vertex and x- and y-intercepts. [Friendly viewing windows can be transformed just as functions can. The quadratic function has been shifted horizontally to the right 2 units. Therefore, add 2 to each x-value in your viewing window. The resulting graph is still not representative since the vertex is not visible. Therefore, change the y-values to –15 and 5. This graph is representative since the vertex, (2, –9), the x-intercepts, (–1, 0) and (5, 0), and the y-intercept, (0, –5), are all visible.]

2. Find a friendly representative viewing window to view the function $h(x) = (x + 2)^2 – 16$. [Subtract 2 from the preset x-values, and set the y-values to –20 and 5. All intercepts and the vertex are visible, and these values can be found exactly using TRACE .]

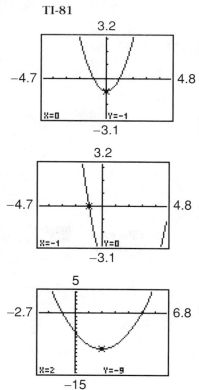

TI-81

GRAPHICS CALCULATOR SKILLS

The Tracing and Zooming Operations

A function such as $f(x) = 0.5x^2 - 1.2x$ is a fairly simple function yet quite worthy of exploration. With its graph displayed, you can obtain a great deal of information about it. Listed here are some observations that are suggested by the representative graph shown.

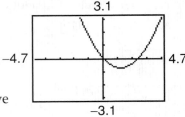

- The graph appears to be an unbroken curve, which may well extend indefinitely.

- The graph appears to have a U shape. The U opens upward and possibly continues in the upward direction indefinitely.

- There are values of x for which the graph is above the x-axis and values of x for which the graph is below the x-axis. There appear to be two values of x for which the graph crosses the x-axis.

- There is an x-value for which the graph has a smallest y-value.

Each of these observations is *qualitative*, that is, each deals with aspects of the graph without regard to precision, numerical inferences, or conclusions.

One of the features of a graphics calculator is its ability to provide numerical information, that is, *quantitative* information. The TRACE feature is one way to approximate, or find, numerical information about the coordinates of points on the graph. Shown in Skill 1 are the steps needed to approximate the coordinates of one of the points where the graph of $f(x) = 0.5x^2 - 1.2x$ crosses the x-axis. Use the friendly viewing window given on page 29.

Skill 1: Approximating Coordinates of Points

Texas Instruments TI-83, TI-82, and TI-81

[TRACE] As you press [◄] or [►], the coordinates of the selected point will be displayed.

Casio CFX-9800G, fx-9700GE, and fx-7700GE

[F1] As you press [◄] or [►], the coordinates of the selected point will be displayed. [FOR THE CFX-9850G, SEE APPENDIX.]

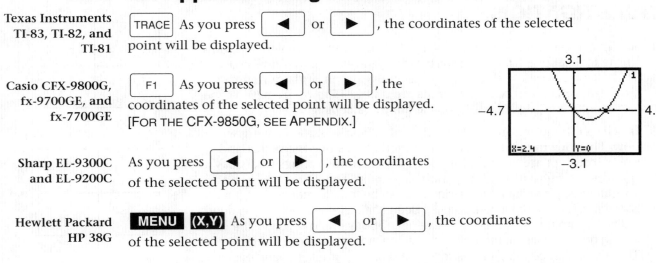

Sharp EL-9300C and EL-9200C

As you press [◄] or [►], the coordinates of the selected point will be displayed.

Hewlett Packard HP 38G

MENU **(X,Y)** As you press [◄] or [►], the coordinates of the selected point will be displayed.

As you trace along the graph from left to right, you will notice the following.

- The y-coordinates are positive and decrease when $x < 0$.

- The y-coordinates are negative and decrease when $0 < x < 1.2$.

- The y-coordinates are negative and increase when $1.2 < x < 2.4$.

- The y-coordinates are positive and increase when $x > 2.4$.

Note: The values of x in this list are exact. When you use the TRACE feature without a friendly viewing window, you may see a value of x like 2.3515789.

The observations made on the preceding page about the graph of $f(x) = 0.5x^2 - 1.2x$ can be explored further with the aid of the ZOOM feature. One of the advantages of the ZOOM feature is that it enables you to take a close-up view of a portion of the graph. The calculator displays shown here and obtained by using the ZOOM feature suggest that the graph of $f(x) = 0.5x^2 - 1.2x$ in the vicinity of $x = -1$ looks like a straight line slanting down. Near $x = 1.2$, the graph looks almost like a horizontal line. In the vicinity of $x = 3$, the graph looks straight but slants up. These diagrams confirm what the tracing operation suggested. To the left of $x = 1.2$, the curve has a downward slant, at $x = 1.2$, the curve is flat, and to the right of $x = 1.2$, the curve has an upward slant.

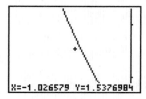 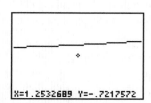 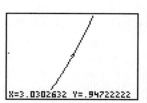

Skill 2 will introduce you to some of the basics involved in the ZOOM operation.

Skill 2: Zooming in on the Graph

Texas Instruments TI-83, TI-82, and TI-81 — With the graph displayed, place the cursor in the vicinity of a portion of the graph you wish to zoom. | ZOOM | **2: Zoom In** | ENTER | ENTER |

Casio CFX-9800G, fx-9700GE, and fx-7700GE — With the graph displayed, place the cursor in the vicinity of a portion of the graph you wish to zoom. | F2 | F3 |
[FOR THE CFX-9850G, SEE APPENDIX.]

Sharp EL-9300C and EL-9200C — With the graph displayed, place the cursor in the vicinity of a portion of the graph you wish to zoom. | ZOOM | **A ZOOM** | **2 IN** | ENTER |

Hewlett Packard HP 38G — With the graph displayed, place the cursor in the vicinity of a portion of the graph you wish to zoom. **MENU** **ZOOM** **In 4×4** **OK**

The friendly viewing windows given on page 29 will often provide a very convenient environment for the tracing feature, since exact values of x can often be found using trace to cursor along a graph. The following descriptions show how to access these special preset windows.

Preset Friendly Viewing Windows

Texas Instruments TI-83 and TI-82 — | ZOOM | **4: ZDecimal** | ENTER | ⟵ $-4.7 \le x \le 4.7$ and $-3.1 \le y \le 3.1$

Casio CFX-9800G and fx-7700GE — | MENU | **6 GRAPH** | EXE | Range | F1 | ⟵ $-4.7 \le x \le 4.7$ and $-3.1 \le y \le 3.1$
[FOR THE CFX-9850G, SEE APPENDIX.]

Casio fx-9700GE — | MENU | **6 GRAPH** | EXE | Range | F1 | ⟵ $-6.3 \le x \le 6.3$ and $-3.706 \le y \le 3.706$

Sharp EL-9300C and EL-9200C — | Range | MENU | **A DEFLT** | ENTER | ⟵ $-4.7 \le x \le 4.7$ and $-3.1 \le y \le 3.1$

Hewlett Packard HP 38G — | | VIEWS LIB | **Decimal** **OK** ⟵ $-6.5 \le x \le 6.5$ and $-3.1 \le y \le 3.2$

INVESTIGATIONS

1. Consider $f(x) = -x^2 + c$, where c is a positive real number.

   a. Choose three different values for c, for example, $c = 1$, 2, and 3. Graph the functions defined by your choices for c. Use a symmetric viewing window and display a representative graph of the set of functions.

   b. Try making some qualitative comparisons and contrasts about the graphs.

   c. Look for inferences you can make about the y-intercept and the x-intercepts of the graph for values of c that get larger and larger.

   d. Then try using the TRACE feature to draw some conclusions that are quantitative. You should discover that the TRACE feature reports the y-intercept of each graph as c or a number very close to it and the x-intercepts of each graph as numbers very close to $\sqrt{c}$ and $-\sqrt{c}$.

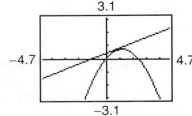

2. The ZOOM operation can help you determine whether two curves cross one another or whether they are close to one another but do not intersect. The graphics calculator display shown here illustrates the graphs of $f(x) = -0.5x^2 + 1.2x$ and $g(x) = 0.4x + 0.4$ in a friendly viewing window defined by $-4.7 \leq x \leq 4.7$ and $-3.1 \leq y \leq 3.1$.

   a. What inferences might one draw from a quick glance at the display? [The line appears to be tangent to the graph of $f(x) = -0.5x^2 + 1.2x$.]

   b. Try exploring what happens when you use the TRACE and ZOOM operations with this pair of functions. You should discover that the graphs do not intersect at all.

   c. To confirm that the graphs do not intersect, try using an algebraic argument, perhaps the quadratic formula, to find any solutions of the equation $-0.5x^2 + 1.2x = 0.4x + 0.4$.

   d. Try investigating this pair of equations: $f(x) = -0.5x^2 + 1.2x$ and $g(x) = 0.4x + 0.3$. [An initial calculator display might suggest that the line is tangent to the curve. In fact, the graphs of these functions intersect in two points.]

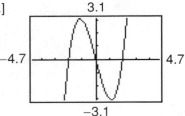

3. Consider $f(x) = (x + 2)x(x - 2)$. Its graph is shown here.

   a. From reading the graph, what do its zeros appear to be?

   b. Try finding out what happens when you use this choice of viewing window in conjunction with the TRACE operation to locate the zeros.

   c. According to the Factor Theorem, the zeros of this function are –2, 0, and 2. Try exploring other ways to use the TRACE command to find zeros of functions.

   d. The calculator display shows a graph that appears to be a sideways S. Try exploring the qualitative and quantitative features of the graph by using the ZOOM command as described in Skills 1 and 2.

4. Choose any linear function you wish, graph it, and trace to any point on the graph. Then ZOOM to view a close-up of the graph. Think about what is true of any close-up view of any straight line.

GRAPHICS CALCULATOR SKILLS

Linear Functions

A *linear function* is any function of the form $f(x) = mx + b$, where m and b are real numbers. The value of m indicates the slope of the line that is the function's graph.

- If $m > 0$, the line slants up to the right. If $m < 0$, the graph slants down to the right. If $m = 0$, the graph is a horizontal line.

The value of b indicates the y-intercept.

- If $b > 0$, the line cuts across the y-axis above the origin. If $b < 0$, the line cuts across the y-axis below the origin. If $b = 0$, the line cuts through the origin.

To graph a function of the form $y = f(x)$, enter function graphing mode, then enter your function. The following key sequences outline the steps needed to graph $f(x) = 1.2x + 2.5$. The friendly viewing windows, found on page 29, will give good results here. On many of the calculators, this convenient window is preset or built in.

Note: Before starting out, clear the calculator of any previous functions and graphs. This is shown in Skill 3.

Skill 1: Graphing a Linear Function

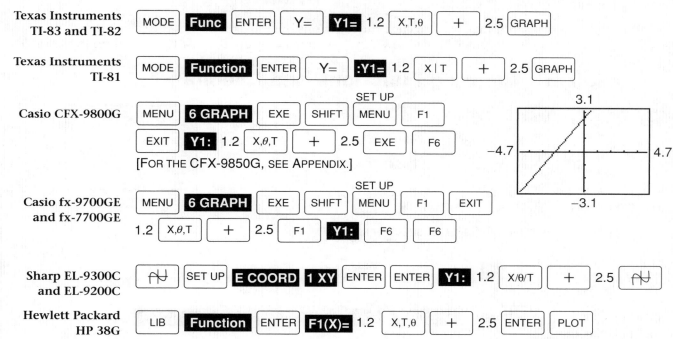

Texas Instruments TI-83 and TI-82
MODE | Func | ENTER | Y= | Y1= | 1.2 | X,T,θ | + | 2.5 | GRAPH

Texas Instruments TI-81
MODE | Function | ENTER | Y= | :Y1= | 1.2 | X | T | + | 2.5 | GRAPH

Casio CFX-9800G
MENU | 6 GRAPH | EXE | SHIFT | MENU (SET UP) | F1
EXIT | Y1: | 1.2 | X,θ,T | + | 2.5 | EXE | F6
[FOR THE CFX-9850G, SEE APPENDIX.]

Casio fx-9700GE and fx-7700GE
MENU | 6 GRAPH | EXE | SHIFT | MENU (SET UP) | F1 | EXIT
1.2 | X,θ,T | + | 2.5 | F1 | Y1: | F6 | F6

Sharp EL-9300C and EL-9200C
⌂ | SET UP | E COORD | 1 XY | ENTER | ENTER | Y1: | 1.2 | X/θ/T | + | 2.5 | ⌂

Hewlett Packard HP 38G
LIB | Function | ENTER | F1(X)= | 1.2 | X,T,θ | + | 2.5 | ENTER | PLOT

Graphing a system of functions, such as system I, will help demonstrate the relationship between the slopes of two lines that are parallel. Graphs of the equations in system II will help demonstrate the relationship between the slopes of two lines that are perpendicular. To explore either system of equations, you will need to enter two functions in the function list. This is shown in Skill 2.

I $\begin{cases} f(x) = 1.2x + 2.5 \\ g(x) = 1.2x - 1.5 \end{cases}$

II $\begin{cases} f(x) = -\dfrac{3}{2}x - 2.5 \\ g(x) = \dfrac{2}{3}x + 1.5 \end{cases}$

Skill 2: Graphing a System of Linear Functions

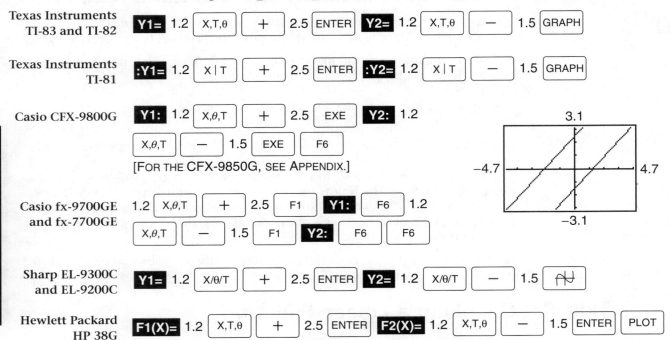

You may wish to clear the functions used in Skill 2 from the calculator. To do this, proceed as follows. (It is assumed they are the first two functions in the list.)

Skill 3: Clearing Functions and Graphs

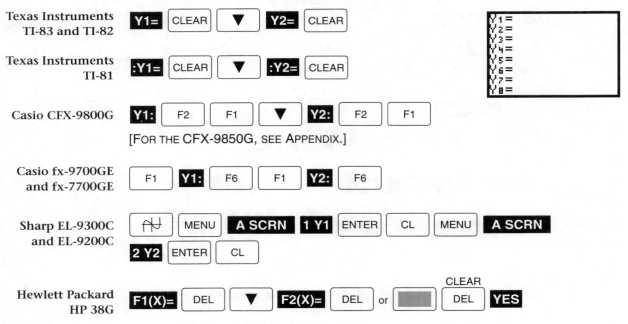

INVESTIGATIONS

1. **a.** Try investigating $f(x) = mx + 1$ and $g(x) = -mx + 1$ for various values of m. Look for a pattern in the graphs. [They are reflections of each other in the y-axis.]

   b. Try exploring $f(x) = 1.5x + b$ and $g(x) = 1.5x - b$ for various values of b. Look for a pattern in the graphs. [One is the translation up (or down) of the other by $|b|$ units.]

With the graphs of the equations displayed, you can identify the solution of the system by moving the cursor to the point where the lines intersect. The coordinates of the location of the cursor are displayed at the bottom of the calculator screen. Using the TRACE feature is illustrated in Skill 2. You can then use the ZOOM feature, also illustrated in Skill 2, to get a close-up view of a small rectangular region in which the solution is located. Use the down arrow to highlight the indicated menu choice such as **2: Zoom In**, and press ENTER to select it.

Skill 2: Tracing and Zooming to the Solution

Texas Instruments TI-83, TI-82, and TI-81
TRACE Use the arrow keys to move to the intersection. Press and select ZOOM **2: Zoom In** ENTER ENTER .

Casio CFX-9800G, fx-9700GE, and fx-7700GE
SHIFT F1 (TRACE) Use the arrow keys to move to the intersection. Press SHIFT F2 F3 . [FOR THE CFX-9850G, SEE APPENDIX.]

Sharp EL-9300C and EL-9200C
Use the arrow keys to move to the intersection. Then press and select ZOOM **A ZOOM** **2 IN** ENTER .

Hewlett Packard HP 38G
Use the arrow keys to move to the intersection. Then press and select **MENU** **ZOOM** **In 4×4** **OK** **(X,Y)** .

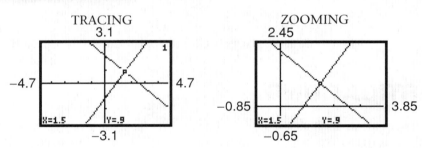

TRACING

ZOOMING

You can observe that the solution is inside the rectangle bounded by $1 \leq x \leq 2$ and $0 \leq y \leq 2$. You can reenter the viewing window menu, reset the ranges for x and y, graph the equations in the new viewing window, and then use TRACE and ZOOM again. There is also a numerical approach, as illustrated in Skill 3.

Skill 3: Using the Solver to Solve the System

Texas Instruments TI-83 and TI-82
With the graph displayed, press 2nd TRACE (CALC) **5: intersect** ENTER . Press ▶ and ◀ to place the cursor near a point of intersection. Press ENTER three times.

Casio CFX-9800G
With the graph displayed, press SHIFT 9 (G-SOLV) F5 . [FOR THE CFX-9850G, SEE APPENDIX.]

Casio fx-9700GE
With the graph displayed, press SHIFT Range (G-SOLV) F5 .

Sharp EL-9300C and EL-9200C
With the graph displayed, press 2nd F EQTN (JUMP) **A NEXT** **1 INTERSEC** ENTER .

Hewlett Packard HP 38G
With the graph displayed, press **MENU** **FCN** **Intersection** **OK** **F2(X)=...** **OK** .

The Casio calculators have a built-in algorithm for solving systems of linear equations. To apply the algorithm to this system, follow the steps in Skill 4. (*Note:* Casio calculators also can be used to solve systems of linear equations that involve three equations and three unknowns. The procedure is the same as that shown here.)

$$\begin{cases} -1.4x + y = -1.2 \\ 1.8x + 2.1y = 4.59 \end{cases}$$

Skill 4: Solving a System Algebraically on the Casio Calculators

Casio CFX-9800G and fx-9700GE

| MENU | **9 EQUA** | EXE | F1 | F1 | (−) | 1.4 | EXE | 1 | EXE | (−) |

| 1.2 | EXE | 1.8 | EXE | 2.1 | EXE | 4.59 | EXE | F1 |

[FOR THE CFX-9850G, SEE APPENDIX.]

Casio fx-7700GE

| MENU | **7 EQUA** | EXE | F1 | (−) | 1.4 | EXE | 1 | EXE | (−) |

| 1.2 | EXE | 1.8 | EXE | 2.1 | EXE | 4.59 | EXE | F1 |

Among the alternatives for solving a system of linear equations is the *row reduction* approach. According to this approach, you multiply one equation in the system by a number and add the result to the other equation. With a wise choice of multiplier, you will eliminate one variable from the second equation. Then divide the second equation by the coefficient of the remaining variable. In similar fashion, you can change the first equation so that it is easily solvable. You can carry out this strategy on the Texas Instruments, Casio, and Sharp calculators.

Finally, you can solve a system of linear equations having as many unknowns as equations by using a matrix product. For more information about the matrix method, see *Matrix Operations and Inverses* (pages 76-78).

INVESTIGATIONS

1. A linear equation such as $y = 0.8x + 2$ has an inverse $y = 1.25x - 2.5$. These equations form a system of two linear equations in two unknowns.

   a. Graph these equations along with the equation $y = x$ in the viewing window whose settings are $-4.7 \le x \le 4.7$ and $-3.1 \le y \le 3.1$.

   b. Look for patterns. You should see that each graph is the reflection of the other in the line $y = x$. Also the system formed by the function and its inverse has a unique solution that lies along the line with equation $y = x$.

   c. Try experimenting with other linear functions with nonzero slope and their inverses to see if every such system is consistent. [The solution is unique if the slope of each line is not 1. If a line has slope 1, the system is consistent but the solution is not unique.]

2. Three lines are *concurrent* in a point if they have exactly one point in common.

   a. Try adding $y = -1.3x - 2$ to the function list started in Skill 1. Look to see if the system, which now contains three equations in two unknowns, has a solution. That is, look to see if the three lines are concurrent.

   b. Try exploring what happens graphically if a system contains at least three linear equations in two unknowns.

3. Try exploring sets of linear functions of the form $y = mx + 1$ for different values of m. The set of such functions is called a *pencil of lines*.

Inequalities

Suppose that you want to explore the inequality $0.8x + 0.6 \geq 2x^2 - 3x - 1$. A fruitful line of thinking is illustrated in this logical string.

$$0.8x + 0.6 \geq 2x^2 - 3x - 1 \rightarrow 0.8x + 0.6 \geq y \geq 2x^2 - 3x - 1 \rightarrow \begin{cases} y \leq 0.8x + 0.6 \\ y \geq 2x^2 - 3x - 1 \end{cases}$$

The graph of $y = 0.8x + 0.6$ is a straight line. The graph of $y = 2x^2 - 3x - 1$ is a parabola. You can now interpret the inequality $0.8x + 0.6 \geq 2x^2 - 3x - 1$ as follows. The solution set is the set of all real numbers x such that the line is above or on the parabola.

Furthermore, the solution of $0.8x + 0.6 \leq 2x^2 - 3x - 1$, an inequality related to the original one, is the set of all real numbers x such that the line is below or on the parabola. To understand this thinking, graph $y = 0.8x + 0.6$ and $y = 2x^2 - 3x - 1$ on a graphics calculator. The procedure is illustrated in Skill 1.

Skill 1: Graphing the Parts of an Inequality

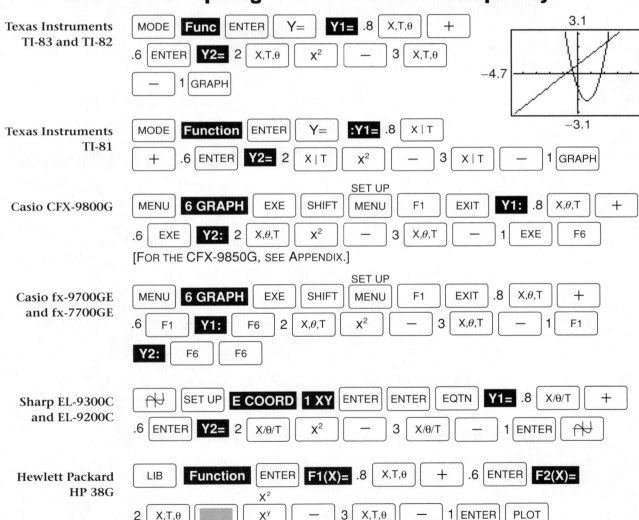

Texas Instruments TI-83 and TI-82: MODE | Func | ENTER | Y= | Y1= | .8 | X,T,θ | + | .6 | ENTER | Y2= | 2 | X,T,θ | x² | − | 3 | X,T,θ | − | 1 | GRAPH

Texas Instruments TI-81: MODE | Function | ENTER | Y= | :Y1= | .8 | X|T | + | .6 | ENTER | Y2= | 2 | X|T | x² | − | 3 | X|T | − | 1 | GRAPH

Casio CFX-9800G: MENU | 6 GRAPH | EXE | SHIFT | MENU[SET UP] | F1 | EXIT | Y1: | .8 | X,θ,T | + | .6 | EXE | Y2: | 2 | X,θ,T | x² | − | 3 | X,θ,T | − | 1 | EXE | F6

[FOR THE CFX-9850G, SEE APPENDIX.]

Casio fx-9700GE and fx-7700GE: MENU | 6 GRAPH | EXE | SHIFT | MENU[SET UP] | F1 | EXIT | .8 | X,θ,T | + | .6 | F1 | Y1: | F6 | 2 | X,θ,T | x² | − | 3 | X,θ,T | − | 1 | F1 | Y2: | F6 | F6

Sharp EL-9300C and EL-9200C: ⌂ | SET UP | E COORD | 1 XY | ENTER | ENTER | EQTN | Y1= | .8 | X/θ/T | + | .6 | ENTER | Y2= | 2 | X/θ/T | x² | − | 3 | X/θ/T | − | 1 | ENTER | ⌂

Hewlett Packard HP 38G: LIB | Function | ENTER | F1(X)= | .8 | X,T,θ | + | .6 | ENTER | F2(X)= | 2 | X,T,θ | [ ] | x^y | x² | − | 3 | X,T,θ | − | 1 | ENTER | PLOT

The diagram that accompanies Skill 1 indicates that the line is above or on the parabola when x is in the interval $a \leq x \leq b$, where a and b are the x-coordinates of the two points in which the graphs intersect. If these two values can be found, the solution of the original inequality can be written down. Now consider various methods for finding a and b.

One method, a graphical approach, is to use the trace feature to approximate a and b. A second method, a numerical approach, available on all the calculators except the TI-81 and Casio fx-7700GE is to use the solve feature. Since the graphs of the functions are already displayed either approach is quite feasible. The solver approach is outlined in Skill 2.

Skill 2: Finding the Coordinates Intersection Points

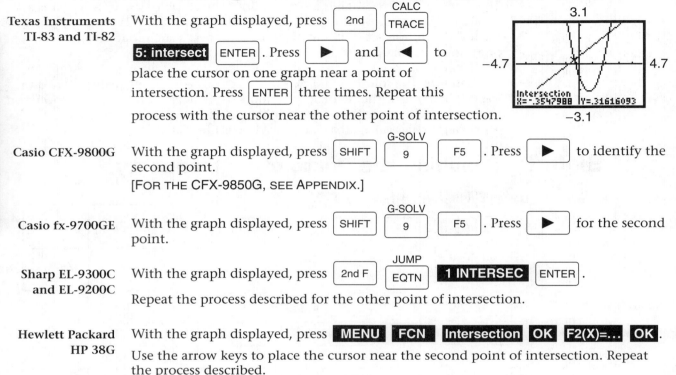

Texas Instruments TI-83 and TI-82

With the graph displayed, press [2nd] [TRACE] (CALC) [5: intersect] [ENTER]. Press [▶] and [◀] to place the cursor on one graph near a point of intersection. Press [ENTER] three times. Repeat this process with the cursor near the other point of intersection.

Casio CFX-9800G

With the graph displayed, press [SHIFT] [9] (G-SOLV) [F5]. Press [▶] to identify the second point.
[FOR THE CFX-9850G, SEE APPENDIX.]

Casio fx-9700GE

With the graph displayed, press [SHIFT] [9] (G-SOLV) [F5]. Press [▶] for the second point.

Sharp EL-9300C and EL-9200C

With the graph displayed, press [2nd F] [EQTN] (JUMP) [1 INTERSEC] [ENTER].
Repeat the process described for the other point of intersection.

Hewlett Packard HP 38G

With the graph displayed, press [MENU] [FCN] [Intersection] [OK] [F2(X)=...] [OK].
Use the arrow keys to place the cursor near the second point of intersection. Repeat the process described.

To a high degree of accuracy, $x \approx -0.3547988$ and $x \approx 2.2547988$. These approximations and the use of $\geq$ in the original inequality give the interval $-0.3547988 \leq x \leq 2.2547988$ as the solution. Furthermore, this information gives $x \leq -0.3547988$ or $x \geq 2.2547988$ as the solution of $0.8x + 0.6 \leq 2x^2 - 3x - 1$.

The inequality $y \geq 2x^2 - 3x - 1$ by itself is an example of an inequality in two variables. Its solution set is a region of the coordinate plane. You can represent its solution by using shading. How to do this is outlined in Skill 3. *Note:* The shade or fill command requires an upper bound for the shading. You may enter any number greater than or equal to the maximum value of y chosen in the viewing window or range settings. Use the down arrow to highlight the indicated menu choice, such as [7: Shade(], and press [ENTER] to select it.

Skill 3: Solving an Inequality in Two Variables

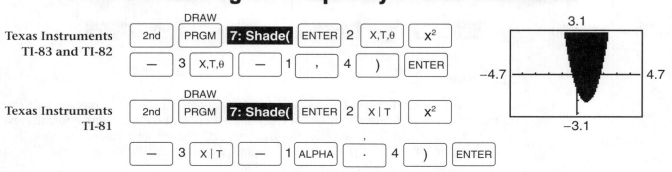

Texas Instruments TI-83 and TI-82

[2nd] [PRGM] (DRAW) [7: Shade(] [ENTER] [2] [X,T,θ] [x²] [—] [3] [X,T,θ] [—] [1] [,] [4] [)] [ENTER]

Texas Instruments TI-81

[2nd] [PRGM] (DRAW) [7: Shade(] [ENTER] [2] [X|T] [x²] [—] [3] [X|T] [—] [1] [ALPHA] [.] [4] [)] [ENTER]

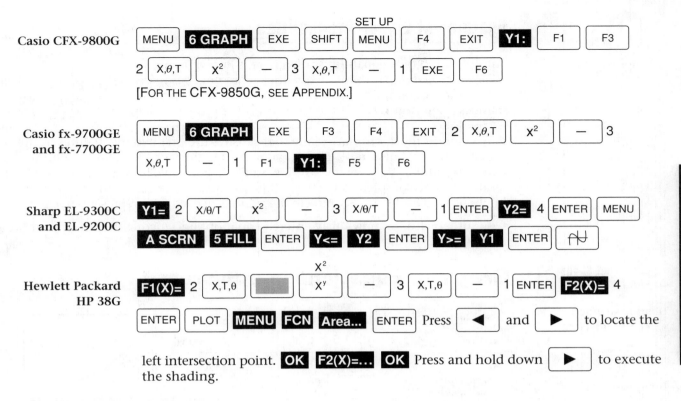

INVESTIGATIONS

1. A local maximum of a function $y = f(x)$ is a point $(c, f(c))$ on the graph of $y = f(x)$ that is above all surrounding points in its vicinity and on the graph.

   a. Consider $f(x) = -0.5x^2 + 2x$ and various constant functions $g(x) = a$. Consider $a = 0, 0.5, 1$, and so on. Find solutions of the inequalities $g(x) \leq f(x)$.

   b. Look for a relationship between the local maximum of $f(x) = -0.5x^2 + 2x$ and the solutions of $g(x) \leq f(x)$ as a increases. [As a increases, the length of the interval that is the solution of $g(x) \leq f(x)$ gets smaller. For some value of a, the length of the solution interval becomes 0. When this happens, you have found the y-coordinate of the local maximum: $a = f(c)$. In other words, the line $g(x) = f(c)$ cuts through the local maximum and the tangent to the graph is horizontal.]

2. A projectile launched from ground level and given an initial velocity v_0 in feet per second has an altitude $f(x)$ in feet x seconds after launch given by the function $f(x) = -16x^2 + v_0x$.

   a. Suppose that the initial velocity is a speed such as 120 ft/s or 180 ft/s. Try exploring, with the use of shading, how to model the time interval over which the altitude of the projectile is greater than a height such as 240 ft or 300 ft.

   b. Look for conclusions you can draw from experimenting with altitudes that are greater and greater.

3. Try using inequalities to investigate this statement: If $y = f(x)$ is a polynomial function, then its graph separates the plane into three regions, the points in the plane that are above the graph, on the graph, and below the graph. Look for ways to validate the statement by exploring different functions.

4. Kylie and Ian incorrectly claimed that $x \leq -3$ is the solution of $-3x - 2 \leq -x + 4$. Try exploring different ways to help them see the error and correct it.

Systems of Linear Inequalities

The collection of inequalities shown here is known as a *system of linear inequalities*. If these inequalities represent *constraints* in a linear programming problem, then the set of points they share as a solution is called the *feasible region* for the problem.

$$\begin{cases} x \geq 0 \\ y \geq 0 \\ y < -0.7x + 2 \\ y < -1.5x + 3 \end{cases}$$

The region in the coordinate plane that satisfies the first two inequalities in the system is the first quadrant. Thus, if you set the minimum for x and the minimum for y at 0 in the viewing window settings, you have begun to represent the solution of the complete collection of inequalities in the system. The maximum values of $x = 3$ and $y = 3.5$ complete the settings needed for a representative graph.

• Be sure to clear all previously entered equations and inequalities from your calculator.

Skill 1: Graphing Equations in a System

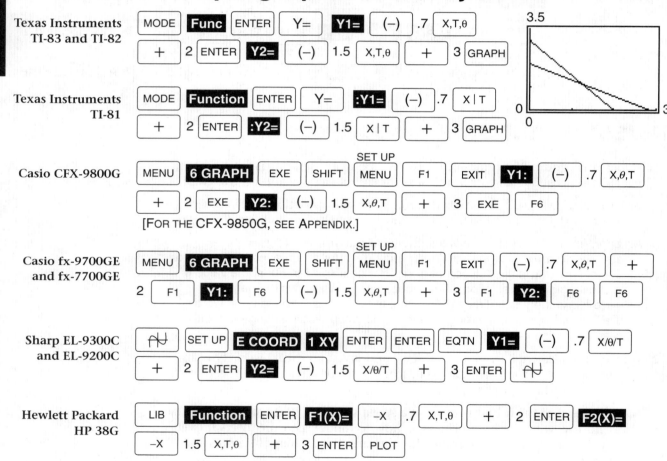

In a linear programming problem, the vertices of the feasible region are important. In the system of linear inequalities discussed here, there are four vertices. The origin (0, 0) is one of them. A second vertex comes from the y-intercept of the equation $f(x) = -0.7x + 2$, the line that is less steep. That vertex is (0, 2).

The third vertex comes from the solution of the system shown here. To find the coordinates of that point of intersection, use the solving routine in your calculator as outlined in Skill 2.

$$\begin{cases} y = -0.7x + 2 \\ y = -1.5x + 3 \end{cases}$$

Skill 2: Finding the Coordinates of Intersection Points

Texas Instruments TI-83 and TI-82   With the graph displayed, press [2nd] [TRACE] (CALC) [5: intersect] [ENTER].

Press [◀] and [▶] to place the cursor on one graph near the point of intersection. Press [ENTER] three times.

Casio CFX-9800G   With the graph displayed, press [SHIFT] [9] (G-SOLV) [F5].
[FOR THE CFX-9850G, SEE APPENDIX.]

Casio fx-9700GE   With the graph displayed, press [SHIFT] [Range] (G-SOLVE) [F5].

Sharp EL-9300C and EL-9200C   With the graph displayed, press [2nd] [EQTN] (JUMP) [1 INTERSEC] [ENTER].

Hewlett Packard HP 38G   With the graph displayed, press [MENU] [FCN] [Intersection] [OK] [F2(X)=...] [OK].

The fourth vertex of the feasible region comes from the x-intercept of $f(x) = -1.5x + 3$, the line that is steeper. That vertex is (2, 0).

Graphics calculators can shade a region whose points have coordinates that satisfy a system of inequalities like this one.
$\begin{cases} y < -(x - 0.5)^2 + 1 \\ y > (x - 0.5)^2 - 1 \end{cases}$

GRAPHICS CALCULATOR SKILLS

Skill 3: Graphing a Shaded Region

Texas Instruments TI-83 and TI-82   [2nd] [PRGM] (DRAW) [7: Shade(] [ENTER] [(] [X,T,θ] [+] [.5] [)] [x²] [−] [1] [,] [(−)] [(] [X,T,θ] [−] [.5] [)] [x²] [+] [1] [)] [ENTER]

Texas Instruments TI-81   [2nd] [PRGM] (DRAW) [7: Shade(] [ENTER] [(] [X|T] [+] [.5] [)] [x²] [−] [1] [ALPHA] [,] [.] [(−)] [(] [X|T] [−] [.5] [)] [x²] [+] [1] [)] [ENTER]

Casio CFX-9800G   [MENU] [6 GRAPH] [EXE] [SHIFT] [MENU] (SET UP) [F4] [EXIT] [Y1:] [F1] [F2] [(−)] [(] [X,θ,T] [−] [.5] [)] [x²] [+] [1] [EXE] [Y2:] [F1] [F1] [(] [X,θ,T] [+] [.5] [)] [x²] [−] [1] [EXE] [F6]
[FOR THE CFX-9850G, SEE APPENDIX.]

Casio fx-9700GE and fx-7700GE   [MENU] [6 GRAPH] [EXE] [F3] [F4] [(−)] [(] [X,θ,T] [−] [.5] [)] [x²] [+] [1] [F1] [Y1:] [F4] [(] [X,θ,T] [+] [.5] [)] [x²] [−] [1] [F1] [Y2:] [F3] [F6]

Sharp EL-9300C and EL-9200C

`Y1=` `(−)` `(` `X/θ/T` `−` `.5` `)` `x²` `+` `1` `ENTER` `Y2=` `(`

`X/θ/T` `+` `.5` `)` `x²` `−` `1` `MENU` `A SCRN` `5 FILL` `ENTER`

`Y<=` `Y1` `ENTER` `Y>=` `Y2` `↲`

Hewlett Packard HP 38G

`F1(X)=` `−x` `(` `X,T,θ` `−` `.5` `)` ▦`x²` `xʸ` `+` `1` `ENTER`

`F2(X)=` `(` `X,T,θ` `+` `.5` `)` ▦`x²` `xʸ` `−` `1` `ENTER` `PLOT`

`MENU` `FCN` `Area...` `ENTER` Press `◄` or `►` to locate the left intersection point. `OK` `F2(X)=...` `OK`. Press and hold down `►` to execute the shading.

INVESTIGATIONS

1. This investigation deals with pairs of absolute value functions. One simple pair is shown here. Notice that one V shape opens down and one V shape opens up.
   $$\begin{cases} y_1(x) \leq -|x| + 2 \\ y_2(x) \geq |x| - 2 \end{cases}$$

   a. Enter $y_1(x) = -|x| + 2$ and $y_2(x) = |x| - 2$. Shade the region between the graphs. Identify the shaded region. [a square]

   b. Try experimenting with variations on the pair of absolute value functions so as to make other quadrilaterals. Try finding out what happens if the graph of the upper V is translated to the left and the lower V is translated the same amount but to the right.

2. This collection of four quadratic functions can be used to construct many different systems of quadratic inequalities. You can use one or more of them at a time.
   $$\begin{cases} y_1(x) = -2x^2 + 2 \\ y_2(x) = -x^2 + 1 \\ y_3(x) = x^2 - 1 \\ y_4(x) = 2x^2 - 2 \end{cases}$$

   a. Try investigating what happens if you make a quadratic system that involves one parabola that opens up and one parabola that opens down.

   b. Investigate what happens when you shade above one and below the other. Then find out what happens when you reverse the order of the two functions in the shade or fill command.

3. a. Consider the functions $f(x) = |x|$ and $g(x) = x^2$. Graph these functions using different viewing window settings. Look for a viewing window that gives a good picture of the graphs near the origin.

   b. Try exploring conditions under which $x^2 > |x|$. (Consider a viewing window defined by $-2 \leq x \leq 2$ and $0 \leq y \leq 6$.) [$x \neq 0$ and $x > 1$ or $x < -1$.]

   c. Consider this pair of equations: $f(x) = a|x|$ and $g(x) = x^2$. Try experimenting with values of a, such as $a = 2, 3$, and so on, to find out whether there is any value of a that makes this statement true: $a|x| > x^2$. [There is no such value of a that will make the statement true.]

   d. Suppose a is a fixed positive number. Use logical thinking to find those values of x for which $a|x| > x^2$. How does your argument help show that no matter how large a is, the graph of $f(x) = a|x|$ will never be entirely above that of $g(x) = x^2$?

4. Try experimenting with how you might use a system of inequalities to model a right triangle whose height is r and whose base is s along with its shaded interior.

GRAPHICS CALCULATOR SKILLS

Quadratic Functions: Graphs, Extrema, and Intersection Points

A *quadratic function* is any function that can be written in the form $f(x) = ax^2 + bx + c$, where a, b, and c are real numbers and a is not 0. One of the distinguishing characteristics of the graph of such a function is its single peak, a *local maximum*, or its single valley, a *local minimum*. The second distinguishing feature of the graph, known as a *parabola*, is that the graph is its own reflection in a vertical line, called the *axis of symmetry*, that passes through the peak or valley, commonly known as the *vertex* of the parabola.

The graph of every quadratic function has either one local maximum or one local minimum. To investigate a quadratic function and its peak or valley, you can begin by graphing the function. The key sequences that follow illustrate how to graph $f(x) = -\frac{1}{2}x^2 + 2x$. *Note:* Since $\frac{1}{2} = .5$, use .5 in place of $\frac{1}{2}$ when entering the expression that defines f.

Skill 1: Graphing a quadratic function

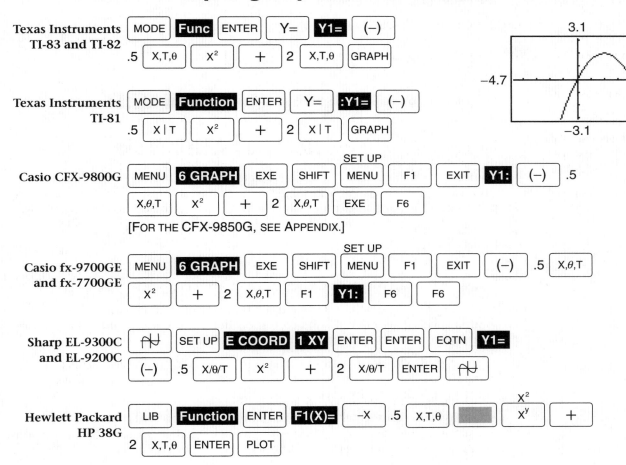

The graph in Skill 1 indicates that the function has a local maximum for some value of x between 1 and 3, perhaps even at $x = 2$. To search for the coordinates of that local maximum, you could use zooming and tracing. There is another option. Various graphics calculators have the built-in capability of delivering the coordinates of a local maximum or local minimum, also called a *local extremum*, of a function. Skill 2 outlines the procedure that will approximate the x-coordinate of local maximum of the graph of $f(x) = -\frac{1}{2}x^2 + 2x$.

Skill 2: Finding the Coordinates of a Local Maximum

Texas Instruments TI-83 and TI-82

With the graph displayed, press

2nd  **4: Maximum**

ENTER . Use the arrow keys to move the cursor to the left of the maximum and press ENTER . Then use ▶ to cursor to the right of the maximum value and press ENTER . Finally, use

◀ to move the cursor close to the peak and between the left and right values you previously selected. Press ENTER . The approximate maximum value will be displayed.

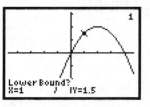

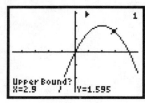

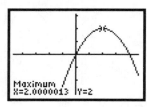

Casio CFX-9800G With the graph displayed, press SHIFT [G-SOLV] 9 F2 .
[FOR THE CFX-9850G, SEE APPENDIX.]

Casio fx-9700GE With the graph displayed, press SHIFT [G-SOLV] Range F2 .

Sharp EL-9300C and EL-9200C With the graph displayed, press 2nd F [JUMP] EQTN **A NEXT** **3 MAX** ENTER .

Hewlett Packard HP 38G With the graph displayed, press **MENU** **FCN** **EXTREMUM** **OK** .

Your graphics calculator can also calculate the coordinates of a minimum of a function such as $f(x) = x^2 - 2x - 2$. The procedure is the same as that used to find the coordinates of a maximum. Instead of choosing the command for maximum, select the command for minimum.

Consider the pair of quadratic functions shown here.

$$\begin{cases} y = -0.2x^2 - 0.5x + 1 \\ y = x^2 - 2x - 2 \end{cases}$$

- You can view the pair as a system of two nonlinear equations in two unknowns. The ordered pairs corresponding to the points of intersection of the graphs are the solutions of the system.

- You can also view the pair of equations as algebraic representations of two parabolas. The coordinates of the points of intersection are the coordinates of the points the graphs have in common.

- Finally, you can view the pair of equations as a representation of a single equation, such as $-0.2x^2 - 0.5x + 1 = x^2 - 2x - 2$. The coordinates of the points of intersection help determine the solution of the equation.

To handle any of the situations just described, follow the procedure outlined in Skill 1 to display the graphs shown here, and then choose the command that gives the coordinates of a point of intersection of two graphs. The appropriate key sequences are shown on the next page.

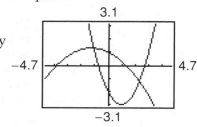

Skill 3: Finding the Coordinates of an Intersection Point

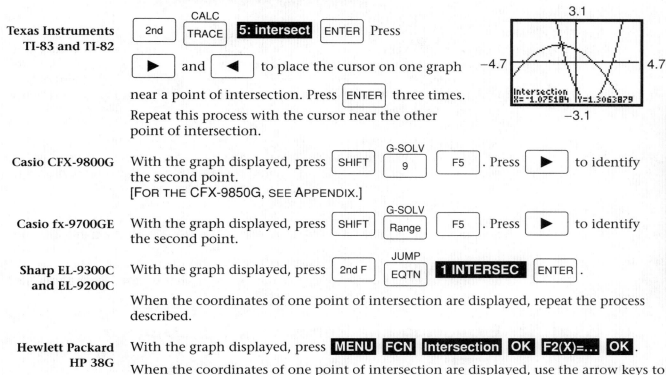

Texas Instruments TI-83 and TI-82

[2nd] [TRACE] (CALC) [5: intersect] [ENTER] Press

[▶] and [◀] to place the cursor on one graph near a point of intersection. Press [ENTER] three times. Repeat this process with the cursor near the other point of intersection.

Casio CFX-9800G

With the graph displayed, press [SHIFT] [9] (G-SOLV) [F5]. Press [▶] to identify the second point.
[FOR THE CFX-9850G, SEE APPENDIX.]

Casio fx-9700GE

With the graph displayed, press [SHIFT] [Range] (G-SOLV) [F5]. Press [▶] to identify the second point.

Sharp EL-9300C and EL-9200C

With the graph displayed, press [2nd F] [EQTN] (JUMP) [1 INTERSEC] [ENTER].

When the coordinates of one point of intersection are displayed, repeat the process described.

Hewlett Packard HP 38G

With the graph displayed, press [MENU] [FCN] [Intersection] [OK] [F2(X)=...] [OK].

When the coordinates of one point of intersection are displayed, use the arrow keys to place the cursor near the second point of intersection. Repeat the process described.

INVESTIGATIONS

1. Use graphing to explore functions of the form $f(x) = ax^2 + bx + c$, where a, b, and c are real numbers and a is not 0.

   a. For each function you examine, find the coordinates of the local maximum or local minimum. Evaluate $-\frac{b}{2a}$ and look for a relationship between that value and the x-coordinate of the local extremum. You should discover that they are equal.

   b. For each function you examine, find the coordinates of the local maximum or local minimum. Compute the value of $b^2 - 4ac$. Look for a relationship between the number of roots of $ax^2 + bx + c = 0$ and the value of $b^2 - 4ac$. Consider the case in which the graph opens upward ($a > 0$) and the case in which the graph opens downward ($a < 0$). You will see that if the graph opens upward and the y-coordinate of the local minimum is negative (positive), the equation has two (no) real roots. If the graph opens downward and the y-coordinate of the local maximum is positive (negative), there are two (no) real roots.

2. Try investigating this claim: If a quadratic function has two real roots, then the local maximum or minimum occurs midway between them. To carry out the investigation, examine quadratic functions whose graphs open upward and quadratic functions whose graphs open downward.

3. Refer to the system shown. Use graphing to illustrate that if $c > 0$, $x = \pm \sqrt{c}$. Look for other patterns and relationships that you can discern from the graphs and the coordinates of the points of intersection.

$$\begin{cases} y = -x^2 + c \\ y = x^2 - c \end{cases}$$

Polynomial Functions and Zeros

A *polynomial function* is any function of the form shown here.

$$y = P(x) = a_n x^n + a_{n-1} x^{n-1} + \ldots + a_1 x^1 + a_0$$

Consider, for example, $x = 0$. The point $(0, P(0))$ is the point at which the graph of $y = P(x)$ crosses the $P(x)$, or y-axis. For this reason, $P(0)$ is called the *y-intercept* of the function. What about those values of x for which $P(x) = 0$? Values of x for which $P(x) = 0$ are called *zeros* of the function $y = P(x)$. The real zeros of the function are called the *x-intercepts* of its graph.

According to the Fundamental Theorem of Algebra, a polynomial function defined by a polynomial of degree n, where n is a positive integer, with complex coefficients has exactly n complex zeros. These zeros may be complex numbers that are not real numbers; they may be real numbers; or some zeros may be complex and some zeros may be real.

Suppose you want to find the zeros of $P(x) = 2x^2 - 6x + 3$. A reasonable first step is to graph the function. Among the possible windows you may choose is the friendly window defined on page 29. The graph of the function is shown in Skill 1. Notice that the graph is a representative graph, that is, it shows the complete behavior of $P(x) = 2x^2 - 6x + 3$.

Skill 1: Exploring the Zeros of a Function

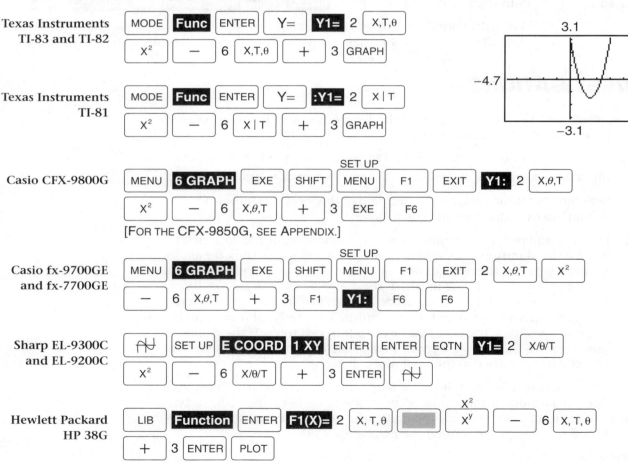

The graph of $P(x) = 2x^2 - 6x + 3$ indicates that the function has two real zeros. One of them is between 0 and 1 and the other one is between 2 and 3. Skill 2 shows how to use the solver in your graphics calculator to approximate them.

Skill 2: Using a Numerical Method to Find Zeros

Enter a guess in the interval $0 \leq x \leq 1$.

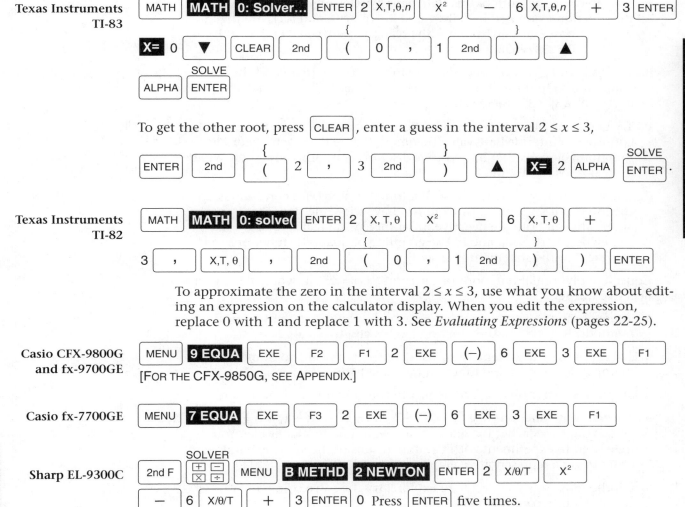

To approximate the zero in the interval $2 \leq x \leq 3$, use what you know about editing an expression on the calculator display. When you edit the expression, replace 0 with 1 and replace 1 with 3. See *Evaluating Expressions* (pages 22-25).

[FOR THE CFX-9850G, SEE APPENDIX.]

To find the zero in the interval $2 \leq x \leq 3$, press ENTER twice. Type 2, then press ENTER three times.

To find the second zero, press NUM 2 ENTER SOLVE.

There are many strategies for finding the roots of an equation or the zeros of a function. Of course, a graphical approach that utilizes the trace feature can be used with friendly windows (see page 29). For more information about solving equations, see *Systems of Linear Functions* (pages 38-40) and *Quadratic Functions: Graphs, Extrema, and Intersection Points* (pages 47-49).

INVESTIGATIONS

1. The *intermediate value theorem* states that if a function $y = f(x)$ is continuous on the interval $a \leq x \leq b$ and $f(a)$ and $f(b)$ are opposite in sign, that is, one has positive value while the other has negative value, then there exists a real zero of $y = f(x)$ somewhere between a and b.

   a. Try exploring the truth of this statement by referring to $P(x) = 2x^2 - 6x + 3$, the function examined in Skills 1 and 2. Find $f(0)$ and $f(1)$. Are the values opposite in sign? What is the real zero between 0 and 1? [$f(0) = 3$; $f(1) = -1$; 0.6339745962]

   b. Find $f(2)$ and $f(3)$. Are they opposite in sign. Use a graph and the solving command to find the real zero of $P(x) = 2x^2 - 6x + 3$ between 2 and 3. [$f(2) = -1$; $f(3) = 3$; 2.366025404]

   c. Try exploring other quadratic functions as suggested in parts **a** and **b** to confirm the intermediate value theorem. Be sure that each quadratic function you construct has positive discriminant, otherwise there will be no real zeros to explore.

   d. The following statement is false. If a function $y = f(x)$ is continuous on the interval $a \leq x \leq b$ and $f(a)$ and $f(b)$ have the same sign, that is, both are positive or both are negative, then there is no real zero of $y = f(x)$ somewhere between a and b. Use graphing of quadratic functions to illustrate that the statement is false.

2. Consider polynomial functions of the form $f_n(x) = x^n - 1$, where n is a nonnegative integer.

   a. Try exploring this statement by using both graphical and numerical methods: If n is even, the function has exactly two real zeros.

   b. Try exploring this statement by using both graphical and numerical methods: If n is odd, the function has exactly one real zero.

3. A function is *even* if $f(-x) = f(x)$ for all x in the domain of the function.

   a. Try graphing functions like $f(x) = x^4 - 2$, $f(x) = x^4 - 2x^2 + 1$, $f(x) = -x^4 + x^2 - 1$, and so on, to see what the graph of an even function looks like. You may need to experiment with the dimensions of the viewing window in order to get a representative graph.

   b. For the functions you decide to explore, try the solve command on the calculator to record the real zeros.

   c. Look for any relationships among the zeros. You should discover that the set of zeros consists of numbers that are additive inverses of one another.

4. Consider these statements. If a polynomial function has even degree, then the number of real zeros must be even. If a polynomial function has odd degree, then the number of real zeros must be odd. Try investigating these statements by examining some of the functions you have already explored. Construct new functions of your own to confirm your observations and conclusions.

5. Consider $f(x) = (x - a)(x - b)$, where a and b are real numbers.

   a. Try exploring these functions for values of a and b that are different and for values of a and b that are equal.

   b. Try drawing conclusions about a quadratic function whose zeros are different, and about a quadratic function whose zeros are the same.

Rational Functions

The graph of every polynomial function is a continuous curve, that is, a curve that has no breaks or jumps in it. If, however, you decide to form a function by dividing one polynomial by another, you will begin to deal with functions that have characteristics quite different from those of polynomial functions.

A *rational function* is a function defined by the quotient of two polynomials. The function $f(x) = \dfrac{x-2}{1.5x+3}$ is a simple example of a rational function. One of the characteristics of the graph of a rational function is the existence of vertical and horizontal *asymptotes*.

The keystrokes that follow illustrate how to graph $f(x) = \dfrac{x-2}{1.5x+3}$.

Skill 1: Graphing a Rational Function

Texas Instruments TI-83 and TI-82

| MODE | **Func** | ENTER | Y= | **Y1=** | (| X,T,θ | − | 2 |) | ÷ |

| (| 1.5 | X,T,θ | + | 3 |) | GRAPH |

Texas Instruments TI-81

| MODE | **Function** | ENTER | Y= | **:Y1=** | (| X | T | − | 2 |) |

| ÷ | (| 1.5 | X | T | + | 3 |) | GRAPH |

Casio CFX-9800G

SET UP

| MENU | **6 GRAPH** | EXE | SHIFT | MENU | F1 | EXIT | **Y1:** | (| X,θ,T | − |

| 2 |) | ÷ | (| 1.5 | X,θ,T | + | 3 |) | EXE | F6 |

[FOR THE CFX-9850G, SEE APPENDIX.]

Casio fx-9700GE and fx-7700GE

SET UP

| MENU | **6 GRAPH** | EXE | SHIFT | MENU | F1 | EXIT | (| Xθ,T | − | 2 |

|) | ÷ | (| 1.5 | X,θ,T | + | 3 |) | F1 | **Y1:** | F6 | F6 |

Sharp EL-9300C and EL-9200C

| ⤶ | SET UP | **E COORD** | **1 XY** | ENTER | ENTER | EQTN | **Y1=** | (| X/θ/T | − |

| 2 |) | ÷ | (| 1.5 | X/θ/T | + | 3 |) | ENTER | ⤶ |

Hewlett Packard HP 38G

| LIB | **Function** | ENTER | **F1(X)=** | (| X,T,θ | − | 2 |) | / |

| (| 1.5 | X,T,θ | + | 3 |) | ENTER | PLOT |

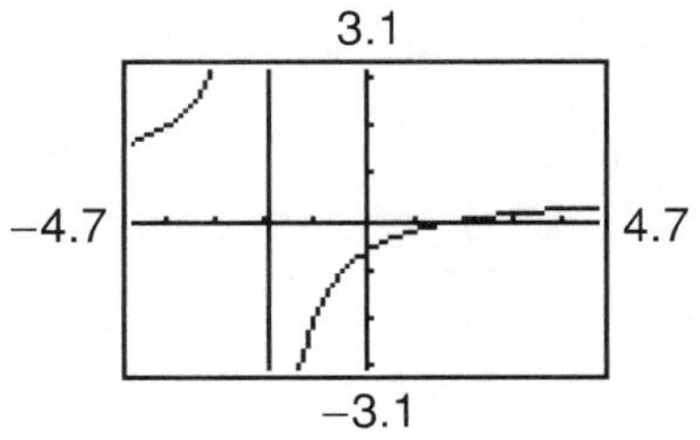

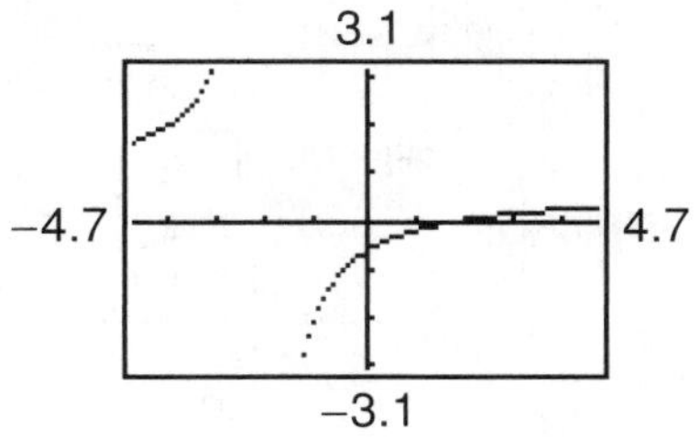

The rational function in Skill 1 has a graph that consists of two distinct branches. One can easily construct rational functions whose graphs consist of three or more branches. Consider the function $f(x) = \dfrac{x-1}{(x+2)x(x-2)}$. When you graph f, you will obtain a calculator display something like this one. As you enter the function into the calculator, be sure to enclose the numerator and the complete denominator in pairs of parentheses.

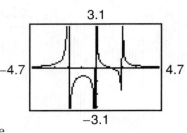

$$(x - 1) \div ((x + 2)x(x - 2))$$

Some rational functions have graphs that almost resemble those of polynomial functions.

For example, if $x \neq -3$, $f(x) = \dfrac{x^2 - 9}{x + 3} = x - 3$. The graph of $f(x) = \dfrac{x^2 - 9}{x + 3}$ is the same as the graph of $y = x - 3$ with a hole in it at $x = -3$. This type of behavior occurs whenever the numerator and the denominator have at least one common factor. Notice from the calculator display that accompanies Skill 2 that the hole is not visible.

Skill 2: Graphing a Function With a Removable Discontinuity

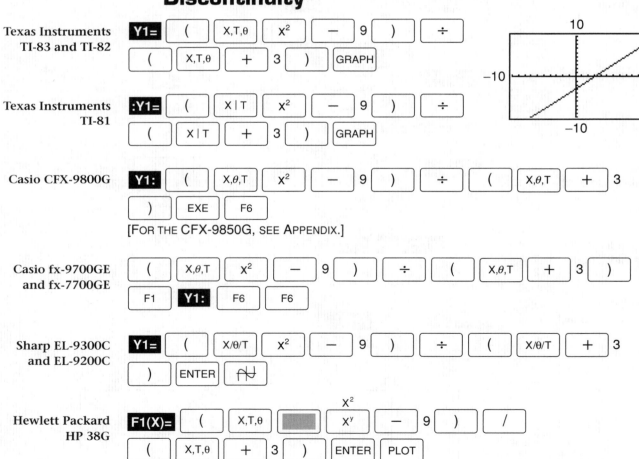

Some rational functions have graphs characterized by an *oblique asymptote*, an asymptote that is neither horizontal nor vertical. Consider the function defined by $f(x) = \dfrac{x^2}{x-1}$. Notice from the display that an oblique, or slant, line segment from (−4, −3) to (4, 5) has been inserted into the graph of f. To show the oblique asymptote for the graph of $f(x) = \dfrac{x^2}{(x-1)}$, graph the function as described in Skill 2. Then follow the appropriate key sequence described in Skill 3. Use the down arrow to highlight the indicated menu choice, such as **2: Line(**, and press ENTER to select it.

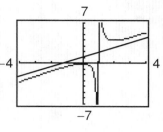

Skill 3: Drawing a Line Segment

Texas Instruments TI-83 and TI-82
With the graph displayed, press 2nd | MODE (QUIT) | 2nd | PRGM (DRAW) | **Draw** | **2: Line(**
ENTER | (−) | 4 | , | (−) | 3 | , | 4 | , | 5 |) | ENTER

Texas Instruments TI-81
With the graph displayed, press 2nd | CLEAR (QUIT) | 2nd | PRGM (DRAW) | **Draw**
2: Line(| ENTER | (−) | 4 | ALPHA | . | (−) | 3 | ALPHA | .
4 | ALPHA | . | 5 |) | ENTER .

Casio CFX-9800G, fx-9700GE, and fx-7700GE
With the graph displayed, press F3 . Use the arrow keys to move the cursor to (−4, −3), press EXE . Move it to (4, 5), press F4 .
[FOR THE CFX-9850G, SEE APPENDIX.]

Sharp EL-9300C and EL-9200C
With the graph displayed, press 2nd F | ZOOM (PLOT) | **C LINE** | **2 FREE** | ENTER .
Use the arrow keys to move the cursor to (−4, −3), press ENTER . Move it to (4, 5), press ENTER .

INVESTIGATIONS

1. **a.** Consider this family of rational functions: $f_c(x) = \dfrac{1}{x-c}$. By graphing particular members of this family, look for a relationship between c and an equation for a vertical asymptote of the graph. [An equation for the vertical asymptote is $x = c$.]

   b. Describe the graph.

2. **a.** Consider $f_a(x) = \dfrac{ax+1}{x-1}$. Graph different members of this family by choosing different values of a, for example, $a = 1, 2, -1,$ and -2.

   b. Try exploring a relationship between a and an equation for a horizontal asymptote of each function you examine. [An equation for the horizontal asymptote is $y = a$.]

3. Graph $f(x) = \dfrac{x^2}{x-1}$ and $g(x) = x + 1$ on the same display. Use the viewing window settings $-4 \le x \le 4$ and $-7 \le y \le 7$. Look for a relationship between the graphs of f and g. [The graph of g might be an oblique asymptote.]

GRAPHICS CALCULATOR SKILLS

Exponential and Logarithmic Functions

An *exponential function* is any function of the form $f(x) = b^x$, where b is a positive real number other than 1. In a particular exponential function, the base b is a fixed number and the exponent x is a variable. For example, if \$1250 is deposited into an account that pays an annual interest rate of 4% compounded yearly, and the money is left untouched for $1\frac{3}{4}$ years, it will grow to $1250(1.04)^{1\frac{3}{4}}$ dollars. Since calculators have parenthesis keys, you need not change $1\frac{3}{4}$ to a decimal to evaluate this exponential expression. The following keystrokes show how the evaluation is made.

Skill 1: Evaluating the Compound Interest Formula

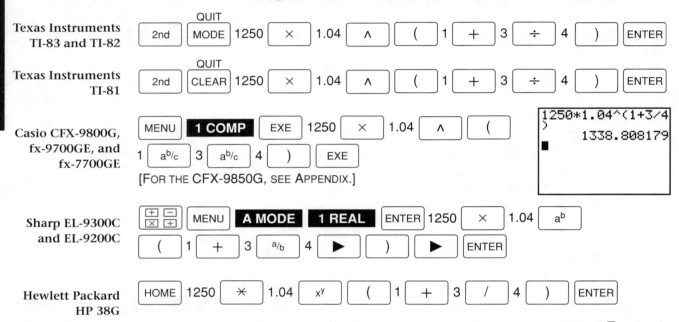

Before the invention of scientific calculators, an expression like $12.25^{\sqrt{3}}$ either would not be evaluated, or it would be evaluated by using powers and roots after the real number was replaced by a rational approximation. Today's calculators make this roundabout and cumbersome procedure obsolete.

Skill 2: Evaluating Expressions with Radicals

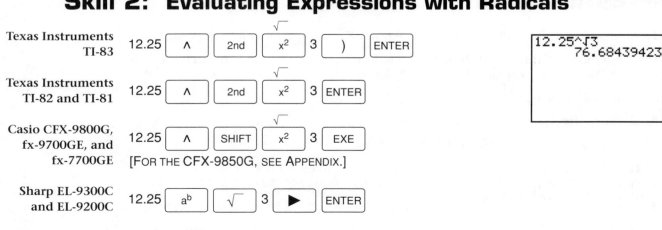

The exponential functions $f(x) = 10^x$ and $f(x) = e^x$ are special exponential functions. The function $f(x) = e^x$ is known as the *natural exponential function*. Because these functions are commonly used, they are built into today's calculators. In the keystrokes that follow, you will see how to evaluate $3.46(10^{1.67})$. To evaluate $3.46e^{1.67}$, press the key involving e^x instead of the key involving 10^x.

Skill 3: Evaluating a Special Exponential Expression

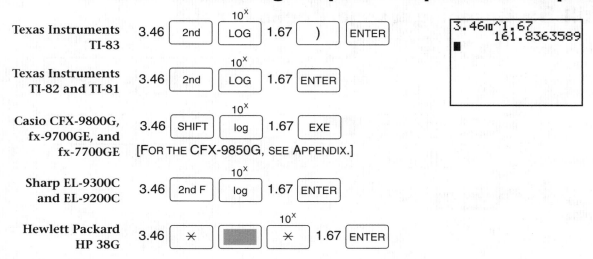

Texas Instruments TI-83
3.46 [2nd] [LOG (10^x)] 1.67 [ )] [ENTER]

Texas Instruments TI-82 and TI-81
3.46 [2nd] [LOG (10^x)] 1.67 [ENTER]

Casio CFX-9800G, fx-9700GE, and fx-7700GE
3.46 [SHIFT] [log (10^x)] 1.67 [EXE]
[FOR THE CFX-9850G, SEE APPENDIX.]

Sharp EL-9300C and EL-9200C
3.46 [2nd F] [log (10^x)] 1.67 [ENTER]

Hewlett Packard HP 38G
3.46 [✳] [▨] [✳ (10^x)] 1.67 [ENTER]

Display:
```
3.46₁₀^1.67
        161.8363589
■
```

Every exponential function $f(x) = b^x$ is one-to-one and therefore has an inverse, called the *logarithmic function with base b*. Logarithmic functions play an important role when it comes to solving exponential equations. Suppose that you deposit $1250 into an account that pays interest annually at the rate of 4%. How long will it take for the initial deposit to grow to $4000? In this problem, time t is the unknown. The equation to be solved is $4000 = 1250(1.04)^t$.

$$4000 = 1250(1.04)^t \longrightarrow t = \frac{\log\left(\frac{4000}{1250}\right)}{\log 1.04}$$

The expression for t is a calculator-ready solution. Skill 4 shows the steps needed to find t.

Skill 4: Solving an Exponential Equation

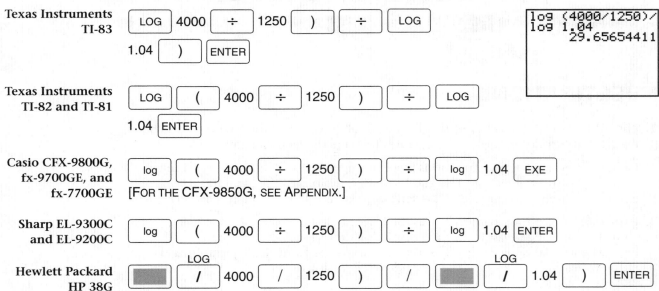

Texas Instruments TI-83
[LOG] 4000 [÷] 1250 [ )] [÷] [LOG] 1.04 [ )] [ENTER]

Texas Instruments TI-82 and TI-81
[LOG] [(] 4000 [÷] 1250 [ )] [÷] [LOG] 1.04 [ENTER]

Casio CFX-9800G, fx-9700GE, and fx-7700GE
[log] [(] 4000 [÷] 1250 [ )] [÷] [log] 1.04 [EXE]
[FOR THE CFX-9850G, SEE APPENDIX.]

Sharp EL-9300C and EL-9200C
[log] [(] 4000 [÷] 1250 [ )] [÷] [log] 1.04 [ENTER]

Hewlett Packard HP 38G
[▨ (LOG) /] 4000 [/] 1250 [ )] [/] [▨ (LOG) /] 1.04 [ )] [ENTER]

Display:
```
log (4000/1250)/
log 1.04
        29.65654411
```

Since every exponential and logarithmic function can be written in the form $y = f(x)$, you graph these types of functions as you would graph any other function of that form. Shown here are the key sequences needed to graph $y = 2 \ln (1.5x + 1)$. Notice that the graph shows the function for the range $-2.7 \le x \le 6.7$ and $-3.1 \le y \le 6.2$. To obtain a graph over these ranges, choose the window or range option and enter these limits when prompted. (See *The Graph Viewing Window* pages 26-29).

Skill 5: Graphing a Logarithmic Function

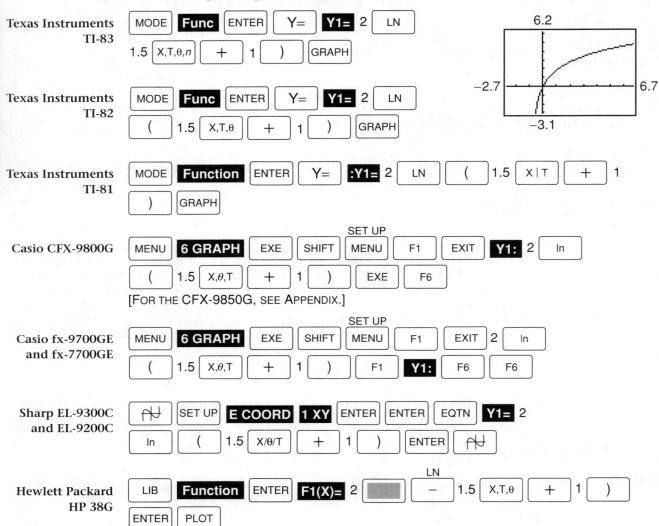

INVESTIGATIONS

1. The display in Skill 4 indicates it will take about 30 years for the $4000 goal to be reached. Try exploring other interest rates to make the goal reachable in less time.

2. Choose three exponential functions by choosing values of b in $f(x) = b^x$. Graph them to test this conjecture: The graphs of all functions of the form $f(x) = b^x$ have exactly one point in common. [The conjecture is true. All such graphs contain (0, 1).]

3. If $\log (ab) = \log a + \log b$, then it must be true that $\log (2b) = \log 2 + \log b$, $\log (3b) = \log 3 + \log b$, $\log (4b) = \log 4 + \log b$, and $\log (5b) = \log 5 + \log b$. Graph $y_1(x) = \log (2x)$ and $y_2(x) = \log 2 + \log x$, to see if their graphs coincide. Try graphing some of the other pairs to confirm that $\log (ab) = \log a + \log b$.

Composites and Inverses of Functions

There are many ways to form new functions from old functions. A rational function is formed by making the quotient of two polynomial functions. Aside from using arithmetic operations to make a new function from other functions, you can compose two functions, that is, create a function whose input value is the output value of another function. The process is called *composition*, and the new function is called a *composite*.

The following key sequences illustrate how you would graphically explore the set of functions shown here. The logical process is as follows.

- Enter $y_1(x) = 1.5x^2 - x$.

- Enter $y_2(x) = 2x - 3$.

- When you enter function $y_2(y_1(x)) = 2(1.5x^2 - x) - 3$, begin to enter the second function again. When you are ready to enter the variable name, recall the first function from the calculator's memory.

$$\begin{cases} y_1(x) = 1.5x^2 - x \\ y_2(x) = 2x - 3 \\ y_3(x) = y_2(y_1(x)) \\ \quad = 2(1.5x^2 - x) - 3 \end{cases}$$

Skill 1: Graphing Functions and Composites

Texas Instruments TI-83

| MODE | **Func** | ENTER | Y= | **Y1=** | 1.5 | X,T,θ |

| X² | — | X,T,θ | ENTER | **Y2=** | 2 | X,T,θ |

| — | 3 | ENTER | **Y3=** | 2 | VARS | **Y-VARS** |

| **1: Function** | ENTER | **1:Y1** | ENTER | — | 3 | GRAPH |

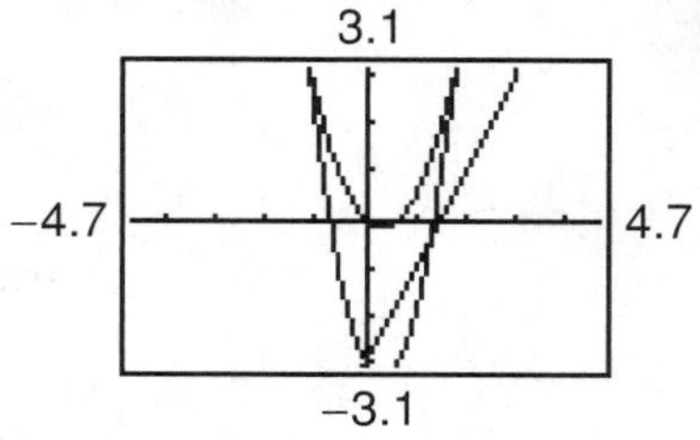

Texas Instruments TI-82

| MODE | **Func** | ENTER | Y= | **Y1=** | 1.5 | X,T,θ | X² | — | X,T,θ | ENTER |

| **Y2=** | 2 | X,T,θ | — | 3 | ENTER | **Y3=** | 2 | 2nd | VARS (Y-VARS) | **1:Function...** | ENTER | **1:Y1** |

| ENTER | — | 3 | GRAPH |

Texas Instruments TI-81

| MODE | **Function** | ENTER | Y= | **:Y1=** | 1.5 | X|T | X² | — | X|T | ENTER |

| **:Y2=** | 2 | X|T | — | 3 | ENTER | **:Y3=** | 2 | 2nd | VARS (Y-VARS) | **Y** | **1:Y1** | ENTER | — | 3 |

| GRAPH |

Casio CFX-9800G

| MENU | **6 GRAPH** | EXE | SHIFT | MENU | F1 | EXIT | **Y1:** | 1.5 | X,θ,T | X² |

| — | X,θ,T | SHIFT | 0 (F-MEM) | F1 | F1 | EXIT | EXE | **Y2:** | 2 | X,θ,T |

| — | 3 | SHIFT | 0 (F-MEM) | F1 | F2 | EXIT | EXE | **Y3:** | 2 | (|

| SHIFT | 0 (F-MEM Z) | F2 | F1 |) | — | 3 | EXE | F6 |

[FOR THE CFX-9850G, SEE APPENDIX.]

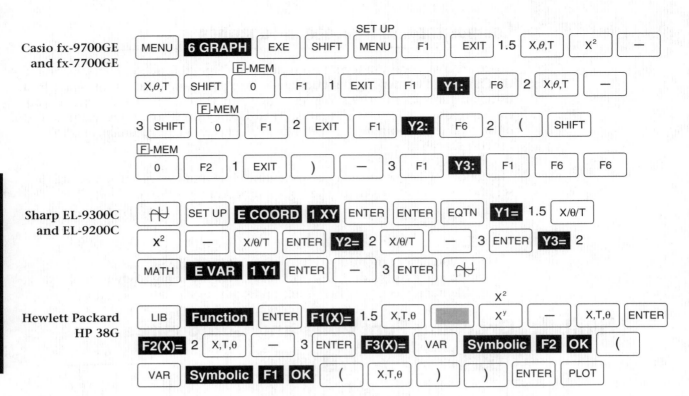

Casio fx-9700GE
and fx-7700GE

Sharp EL-9300C
and EL-9200C

Hewlett Packard
HP 38G

The TI-83 and TI-82 graphics calculators allow you to enter a function and automatically graph it and its inverse. This feature is particularly useful since it allows you to visualize a function and its inverse without actually writing an equation for the inverse, and whether or not the inverse is a function. Skill 2 shows how to graph $f(x) = 0.25x^2 - 1$ and its inverse. Use the down arrow to highlight the indicated menu choice, such as **8: DrawInv** , and press ENTER to select it.

Skill 2: Graphing a Function and its Inverse on the TI-83 and TI-82

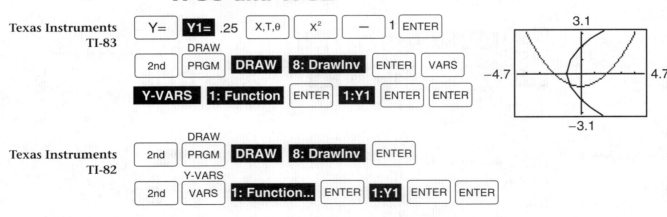

Texas Instruments
TI-83

Texas Instruments
TI-82

On the Casio, Sharp, and Hewlett Packard calculators, you can graph a linear function and its inverse by using the equations shown. You will need to supply the values of m and b.

$$\begin{cases} f(x) = mx + b \\ f^{-1}(x) = \dfrac{x - b}{m} \end{cases}$$

To make the relationship between a function and its inverse clear, you can add the function $f(x) = x$ to the function list in the calculator. When all three graphs are displayed, it will be clear that a function and its inverse are reflections of each other in the line $y = x$. For more information about functions and inverses of them, see *Parametric Equations* (pages 66-68).

INVESTIGATIONS

1. What happens when a linear function is composed with another linear function? What must be true of a composite of them? To find the answers to these questions, let $f(x) = 2x + 1$ and $g(x) = 0.3x - 2$.

   a. Graph $f(x) = 2x + 1$ and $g(x) = 0.3x - 2$ on the same calculator display.

   b. To the function list in the calculator, add the functions $y = f(g(x))$ and $y = g(f(x))$. Graph all the functions in the list. You should see that all the graphs on the display are straight lines. This helps answer both questions posed. The composition of these two linear functions is another linear function.

   c. Try experimenting with other linear functions to confirm this fact.

   d. Begin with a different pair of linear functions to see if the observation made in part c remains true. This time, let $f(x) = 0.8x + 0.5$ and $g(x) = 1.4x - 0.7$.

   e. Repeat parts b and c. The same thing happens. Again there are four straight lines on the calculator display. No matter what two linear functions are chosen, the composite of them will be another linear function.

2. This activity deals with the family of functions $f(x) = \frac{k}{x}$.

   a. Choose three nonzero values of k and enter the functions your values of k define.

   b. To the function list in the calculator, add $g(x) = x$. Graph all four functions in the list on the same calculator display. Use the viewing window $-4.7 \leq x \leq 4.7$ and $-3.1 \leq y \leq 3.1$.

   c. The calculator display shows a set of six symmetric curves and one straight line. What does this say about the functions graphed in part b? [Since the pair of curves corresponding to a particular value of k is symmetric about the line $y = x$, each function of the form $f(x) = \frac{k}{x}$ is its own inverse.]

3. The functions $f(x) = e^x$ and $g(x) = \ln x$ are built into each graphics calculator.

   a. In the graph viewing window, set the ranges for x and y at $-4.7 \leq x \leq 4.7$ and $-3.1 \leq y \leq 3.1$. Enter $f(x) = e^x$ and $g(x) = \ln x$ into the calculator's function list. Also enter the function $y = x$. Graph them and look for a visual relationship between $f(x) = e^x$ and $g(x) = \ln x$ from the display. The display should confirm that $f(x) = e^x$ and $g(x) = \ln x$ are inverses of one another.

   b. To the function list started in part a, add the composite of f with g and g with f. Look for conclusions that can be drawn from the fact that the function list contains five functions but the display shows only three graphs. One conclusion you can draw is that both composites are equal and their graphs are the line $y = x$. This also means that $f(x) = e^x$ and $g(x) = \ln x$ are inverses of each other.

4. The functions $f(x) = x^2$ and $g(x) = \sqrt{x}$ are built into each graphics calculator. Consider the viewing window $0 \leq x \leq 9$ and $0 \leq y \leq 6$. Try investigating different ways to illustrate that if $x > 0$, these functions are inverses of one another.

5. You can begin with a single function and compose it with itself over and over again. This process is known as *iteration*. Suppose that you begin with $f(x) = x + 0.25$. Try making a function list that contains f, f composed with itself, f composed with the composite, and so on. Try studying the behavior of the iteration process for this function. Try exploring different functions whose iteration process carries out successive translations downward.

GRAPHICS CALCULATOR SKILLS

Trigonometric Functions

The sine and cosine functions are the two basic *trigonometric functions*. The evaluation of an expression involving the sine or cosine will arise in the process of solving a problem using the law of sines or the law of cosines. For example, this diagram shows $\triangle ABC$ with certain parts known and certain parts unknown. According to the law of sines, the equation $\frac{x}{\sin 32.6°} = \frac{20}{\sin 47.2°}$ can be used to find x.

A calculator-ready solution for x is $x = \frac{20 \sin 32.6°}{\sin 47.2°}$. The following key sequences illustrate the evaluation of this expression, and thus, the solution of the problem.

Skill 1: Evaluating a Trigonometric Expression

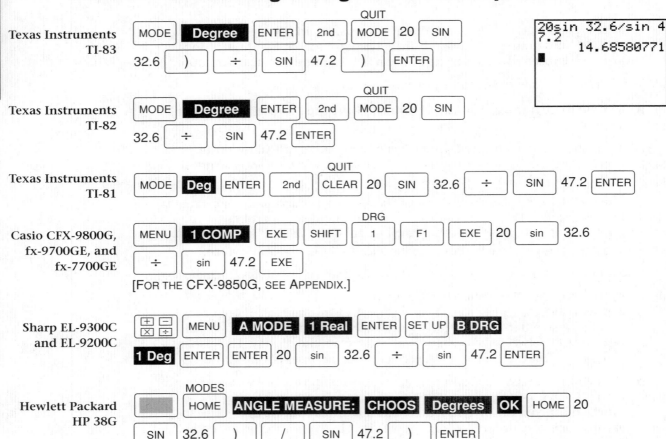

Texas Instruments TI-83

Texas Instruments TI-82

Texas Instruments TI-81

Casio CFX-9800G, fx-9700GE, and fx-7700GE
[FOR THE CFX-9850G, SEE APPENDIX.]

Sharp EL-9300C and EL-9200C

Hewlett Packard HP 38G

In many problems, the goal is to find the measure of an angle, in either degrees or radians, for which the value of a trigonometric function is given. The steps outlined here show how to find x, in degrees, if $x = \cos^{-1}(-0.6542)$.

Skill 2: Finding an Angle Measure

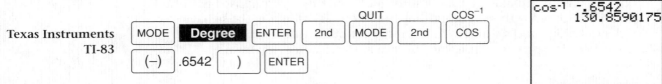

Texas Instruments TI-83

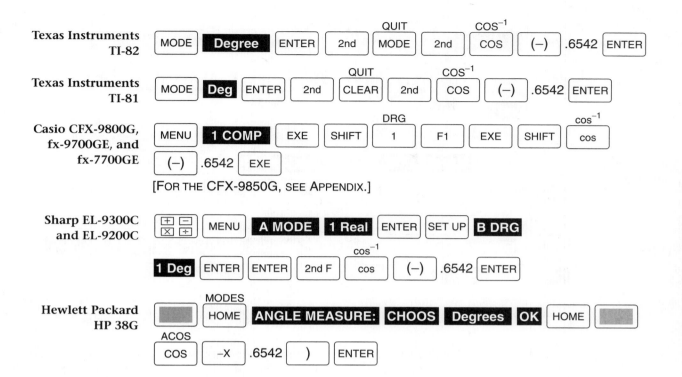

Activities that involve the graphing of trigonometric functions can be very beneficial. Since each trigonometric function has the form $y = f(x)$, graph these functions as you would graph any other function of that form. One consideration of importance is the selection of degree or radian mode. If you wish to graph a function over the interval $-360° \leq x \leq 360°$, select degree mode. If the domain is $-2\pi \leq x \leq 2\pi$, select radian mode. After entering graphing mode, make the appropriate changes to the viewing window settings.

Follow the appropriate key sequence that follows to graph $f(x) = 2.5 \sin (x - 90°)$. In the diagram, the viewing window settings are $-360° \leq x \leq 360°$ and $-3 \leq y \leq 3$.

Skill 3: Graphing a Function Involving Sine

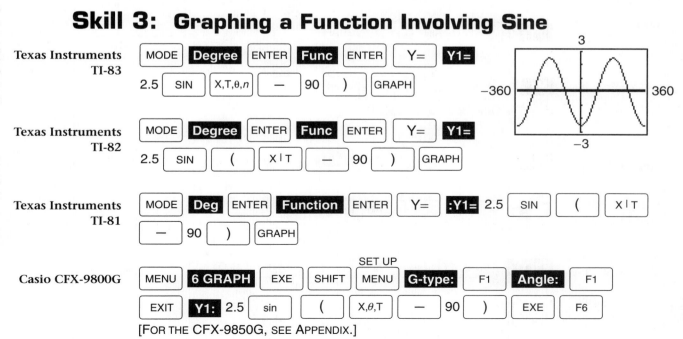

[FOR THE CFX-9850G, SEE APPENDIX.]

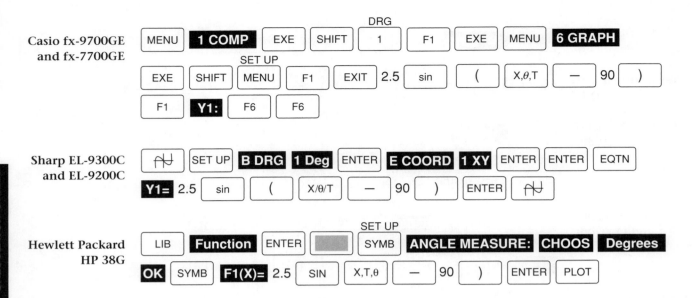

Casio fx-9700GE and fx-7700GE

Sharp EL-9300C and EL-9200C

Hewlett Packard HP 38G

Graphics calculators have keys that enable you to graph the inverse of a trigonometric function, such as $f(x) = 2 \sin^{-1} x$. When setting values for the viewing window, remember that the domain for the inverse sine is $-1 \leq x \leq 1$. The procedure for graphing $f(x) = 2 \sin^{-1} x$ is outlined in what follows. First, enter function graphing mode as previously described.

Skill 4: Graphing a Trigonometric Inverse

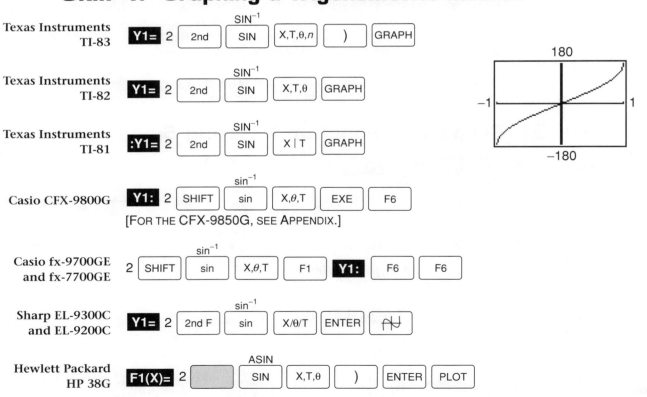

INVESTIGATIONS

1. This investigation deals with variations on the function $f_1(x) = \sin x$.

   a. Consider $f_2(x) = a \sin x$, where a is a nonzero real number. Try investigating the relationship between the graph of $f_2(x) = a \sin x$ and that of $f_1(x) = \sin x$ for different choices of a greater than 1 and choices of a between 0 and 1. For example, consider $f_2(x) = 2 \sin x$ and $f_2(x) = 0.5 \sin x$.

   b. Consider $f_2(x) = \sin (bx)$, where b is a nonzero real number. Try investigating the relationship between the graph of $f_2(x) = \sin (bx)$ and that of $f_1(x) = \sin x$ for different choices of b greater than 1 and choices of b between 0 and 1. For example, consider $f_2(x) = \sin (2x)$ and $f_2(x) = \sin (0.5x)$.

   c. Consider $f_2(x) = \sin x + c$, where c is a nonzero real number. Try investigating the relationship between the graph of $f_2(x) = \sin x + c$ and that of $f_1(x) = \sin x$ for different choices of c that are positive and negative. For example, consider $f_2(x) = \sin x + 1$ and $f_2(x) = \sin x - 1$.

   d. Lastly, consider $f_2(x) = \sin (x + d)$, where d is a nonzero real number. Try investigating the relationship between the graph of $f_2(x) = \sin (x + d)$ and that of $f_1(x) = \sin x$ for different choices of d that are positive and negative. For example, consider $f_2(x) = \sin (x + 90°)$ and $f_2(x) = \sin (x - 90°)$.

2. This investigation deals with equations known as *identities*, equations true for all values of the variable(s) for which both sides of the equation are defined.

   a. If it is true that $\tan x = \frac{\sin x}{\cos x}$ for all values of x for which both sides of the equation are defined, then you would expect the graphs of the functions defined by each side of the equation to be the same. Verify that this is so by graphing $f_1(x) = \tan x$ and $f_2(x) = \frac{\sin x}{\cos x}$ from the same function list.

   b. Use a graphics calculator to make a plausible argument that the equation $\sin (x + 90°) = \cos x$ is an identity.

   c. The equation $\sin (x + 90°) = \cos x$ is an example of an identity and also an example of a horizontal translation. Look for other relationships between the sine function and the cosine function that arise from some horizontal translation of the sine or cosine function.

3. This investigation deals with solving triangles. A triangle, $\triangle ABC$, is shown here. Each side is labeled with the lowercase letter corresponding to the label of the vertex opposite it.

   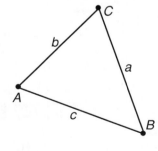

   a. Suppose that the measures of $\angle A$ and $\angle C$ along with the length b of $\overline{AC}$ are known. Suppose you want to find a. (This problem is known as an ASA problem (angle-side-angle.)) Write a formula that will give a in terms of the known information.

   $$\left[\frac{a}{\sin A} = \frac{b}{\sin B} ; \quad a = \frac{b \sin A}{\sin B} ; \quad a = \frac{b \sin A}{\sin (180° - (A + C))} \right]$$

   b. Look for ways to use the formula from part a along with the skill of editing a formula on your calculator to solve related ASA problems in which the length of one side is the unknown. For example, consider finding a given that $\angle A = 60°$, $\angle C = 45°$, and $b = 10$, then $\angle A = 60°$, $\angle C = 45°$, and $b = 12$, and so on.

GRAPHICS CALCULATOR SKILLS

Parametric Equations

A curve C in the plane is defined parametrically if the x- and y-coordinates of any point on the curve are defined in terms of a third variable, such as t. The variable t is called a *parameter*. One advantage of the use of parametric equations is that the location $P(x, y) = P(x(t), y(t))$ of a point moving in space can be studied as a function of elapsed time t after an initial point in time t_0. A second advantage is that many curves that do not represent functions can be modeled algebraically.

$$C \longrightarrow C(x, y) \text{ such that } \begin{cases} x = x(t) \\ y = y(t) \end{cases}$$

To graph a pair of parametric equations, you first must be in parametric mode. After entering parametric mode, proceed much the same way as you would when graphing a function of the form $y = f(x)$.

To graph the curve in the plane defined parametrically by these equations, follow the appropriate key sequence described here.

$$\begin{cases} x = 2t + 1 \\ y = 0.5t^2 - 1 \\ -7 \le t \le 7 \end{cases}$$

Skill 1: Graphing Algebraic Parametric Equations

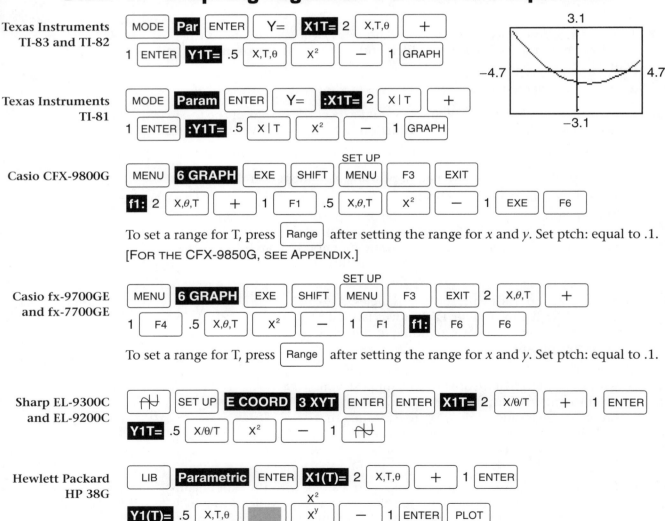

Texas Instruments TI-83 and TI-82

MODE **Par** ENTER Y= **X1T=** 2 X,T,θ + 1 ENTER **Y1T=** .5 X,T,θ X² − 1 GRAPH

Texas Instruments TI-81

MODE **Param** ENTER Y= **:X1T=** 2 X|T + 1 ENTER **:Y1T=** .5 X|T X² − 1 GRAPH

Casio CFX-9800G

MENU **6 GRAPH** EXE SHIFT MENU (SET UP) F3 EXIT
f1: 2 X,θ,T + 1 F1 .5 X,θ,T X² − 1 EXE F6

To set a range for T, press Range after setting the range for x and y. Set ptch: equal to .1. [FOR THE CFX-9850G, SEE APPENDIX.]

Casio fx-9700GE and fx-7700GE

MENU **6 GRAPH** EXE SHIFT MENU (SET UP) F3 EXIT 2 X,θ,T + 1 F4 .5 X,θ,T X² − 1 F1 **f1:** F6 F6

To set a range for T, press Range after setting the range for x and y. Set ptch: equal to .1.

Sharp EL-9300C and EL-9200C

◄┘ SET UP **E COORD** **3 XYT** ENTER ENTER **X1T=** 2 X/θ/T + 1 ENTER **Y1T=** .5 X/θ/T X² − 1 ◄┘

Hewlett Packard HP 38G

LIB **Parametric** ENTER **X1(T)=** 2 X,T,θ + 1 ENTER **Y1(T)=** .5 X,T,θ [ ] X^y − 1 ENTER PLOT

When graphing a curve defined by a pair of trigonometric functions such as the pair shown here, keep in mind the need to select either degrees or radians before graphing the pair of equations.

$$\begin{cases} x = 2.5 \cos (1.5t + 2.1) \\ y = 2.4 \sin (3.5t + 3) \end{cases}$$

- Make sure that you have chosen parametric mode on your calculator.

- Select either degree or radian measure, whichever is appropriate. In the keystrokes that follow, radian mode was selected and a preset viewing window was chosen.

Skill 2: Graphing Trigonometric Parametric Equations

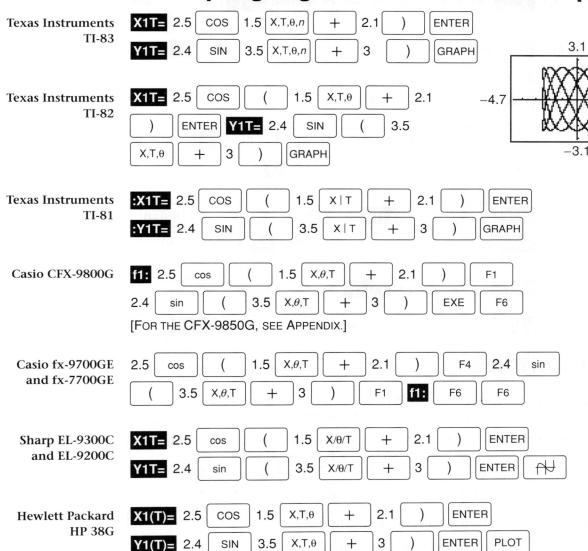

Many functions have inverses but many functions do not. You can tell graphically if a function has an inverse by graphing the function and graphing horizontal lines in the plane. If any horizontal line passes through the graph in more than one point, the function does not have an inverse. If no horizontal line passes through the graph more than once, the function does have an inverse. This test is known as the *horizontal line test*. If $y = f(x)$ has an inverse and (x, y) satisfies $y = f(x)$, then (y, x) satisfies its inverse. This means that the inverse of a function is a reflection of the graph of the function in the line $y = x$. The key sequences that follow illustrate how to use parametric equations to graph $y = 0.5x^2 - 1$ and its inverse.

The keystrokes shown will help you explore the recursive sequence $c_n = (1.04)c_{n-1}$.

Skill 2: Exploring a Recursive Sequence

Texas Instruments TI-83

[MODE] [Seq] [ENTER] [Dot] [ENTER] [WINDOW] ... [ENTER] [Y=]

[Un=] 1.04 [X,T,θ,n] [2nd] [GRAPH]   (TABLE above GRAPH)

n	Un
0	1000
1	1040
2	1081.6
3	1124.9
4	1169.9
5	1216.7
6	1265.3

$n=0$

To display the graph, press [GRAPH].

Texas Instruments TI-82

[MODE] [Seq] [ENTER] [Dot] [ENTER] [WINDOW] ... [ENTER] [Y=]

[Un=] 1.04 [2nd] [7] [2nd] [GRAPH]   (U_{n-1} above 7; TABLE above GRAPH)   To display the graph, press [GRAPH].

Hewlett Packard HP 38G

[LIB] [Sequence] [ENTER] [▓] [NUM] [NUMSTART] 1 [ENTER]   (SETUP above)

[NUMSTEP] 1 [ENTER] [SYMB] [U1(1)=] 1000 [ENTER] [U1(2)=] 1040

[ENTER] [U1(N)=] 1.04 [U1] [(N−1)] [ENTER] [NUM]   To display the graph, press [PLOT].

A *series* is the sum you get when you add the terms of a sequence. As with sequences, a series can be defined explicitly or recursively. This pair of formulas tells you how much would be in an account that pays 4% interest compounded annually if you deposit $1000 at the beginning of each year. A financial arrangement in which you deposit a fixed amount into an account at regular intervals is called an *annuity*.

$$\begin{cases} A_n = 1000 \left(\dfrac{1 - 1.04^n}{1 - 1.04} \right), \text{ where } n \geq 1 \\ A_1 = 1000 \text{ and } A_n = 1000 + 0.04A_{n-1}, \text{ where } n \geq 2 \end{cases}$$

INVESTIGATIONS

1. a. The sequence $c_n = 1000(1.04)^n$ is an example of an increasing exponential sequence. Try exploring how you might change the sequence so that it is a decreasing exponential sequence. [Replace 1.04 with a positive number less than 1.]

   b. Try exploring how you might change the sequence so that it is an increasing exponential sequence that contains the origin. [Replace $1000(1.04)^n$ with any downward translation that is at least one unit.]

   c. Consider how the sequences $c_n = 1000(1.04)^n$ and $d_n = 1000(1.05)^n$ compare. These sequences represent the amount a depositor would have if the interest rate is 4% and 5%.

   d. Consider how the sequences $e_n = 1000(1.04)^n$ and $f_n = 500(1.05)^n$ compare. These sequences represent the amount a depositor would have if the interest rate is 4% and 5% given an inital investment of $1000 and $500.

2. The sequence $u_1 = 1$, $u_2 = 1$, and $u_n = u_{n-1} + u_{n-2}$, where n is a positive integer 3 or more, is known as the *Fibonacci sequence*.

   a. Try exploring patterns in the Fibonacci sequence.

   b. Make a new sequence $r_n = \dfrac{u_n}{u_{n-1}}$. Try exploring it for large values of n and the value of $\dfrac{-1 + \sqrt{5}}{2}$. Then explore $R_n = \dfrac{1}{\frac{u_n}{u_{n-1}}}$ and $\dfrac{1 + \sqrt{5}}{2}$.

GRAPHICS CALCULATOR SKILLS

Second-Degree Equations in Two Variables

Every polynomial, exponential, logarithmic, and trigonometric function is a function defined *explicitly*. That is, each function has the form $y = f(x)$. You can define functions *implicitly*. The three equations shown define functions implicitly.

ellipse: $4x^2 + 9y^2 = 36$ circle: $x^2 + y^2 = 9$ hyperbola: $4x^2 - 9y^2 = 36$

To explore ellipses, circles, and hyperbolas with a graphics calculator, you will need to split the implicit equations into two explicit equations. Shown here is the process applied to $4x^2 + 9y^2 = 36$.

$$4x^2 + 9y^2 = 36$$
$$9y^2 = 36 - 4x^2$$

Once the split of one equation into two equations has been made, you can enter the two equations into the calculator.

$$y^2 = \frac{36 - 4x^2}{9}$$

The final pair of equations shown is itself informative.

- x can only take on values in the interval $-3 \le x \le 3$.

- The complete curve is a curve in the first and second quadrants along with its reflection in the x-axis.

$$\begin{cases} y = \sqrt{\dfrac{36 - 4x^2}{9}} \\ y = -\sqrt{\dfrac{36 - 4x^2}{9}} \end{cases}$$

Skill 1: Graphing an Ellipse

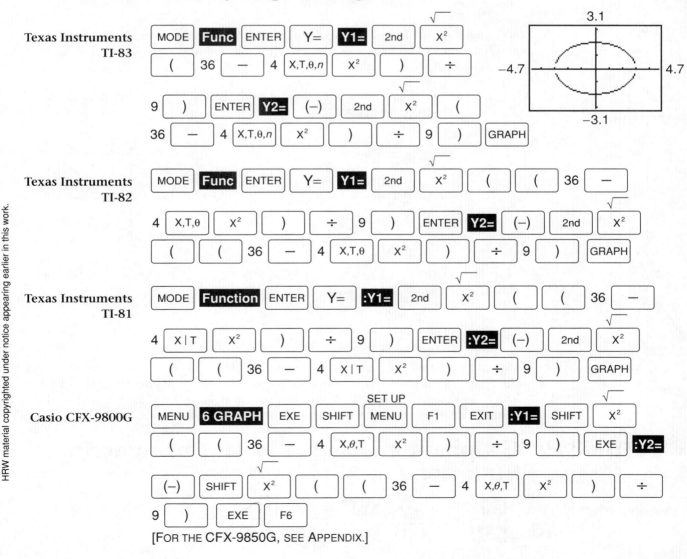

Texas Instruments TI-83

Texas Instruments TI-82

Texas Instruments TI-81

Casio CFX-9800G

[FOR THE CFX-9850G, SEE APPENDIX.]

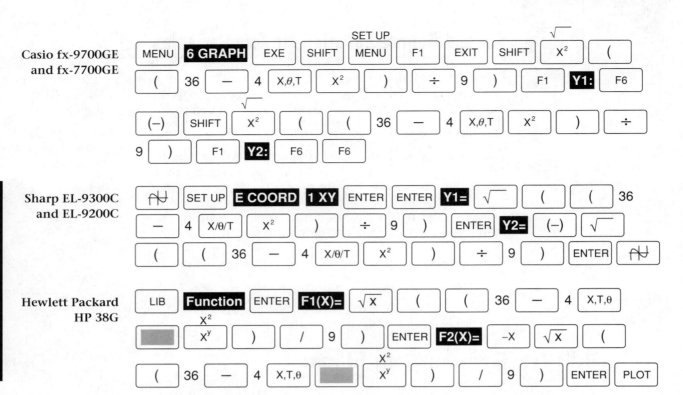

Casio fx-9700GE and fx-7700GE

Sharp EL-9300C and EL-9200C

Hewlett Packard HP 38G

You might notice from the calculator display that accompanies Skill 1 that the graph appears to be broken in the vicinity of $x = -3$ and $x = 3$. This behavior is observed with some window settings.

The following displays illustrate the graphs of the circle and hyperbola introduced on the preceding page.

$x^2 + y^2 = 9$

$4x^2 - 9y^2 = 36$

Note: The circle looks like a true circle, and not an ellipse. This is one of the advantages of using a square viewing window, as are the preset viewing windows discussed on page 29.

One of the useful applications of parametric equations is the representation of second-degree equations in two variables. The following keystrokes illustrate how to graph $4x^2 + 9y^2 = 36$ using the parametric equations shown here. Use radian mode.

$$\begin{cases} x(t) = 3 \cos t \\ y(t) = 2 \sin t \\ 0 \le t \le 2\pi \end{cases}$$

Skill 2: Graphing an Ellipse With Parametric Equations

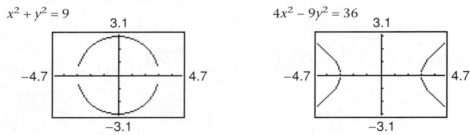

Texas Instruments TI-83

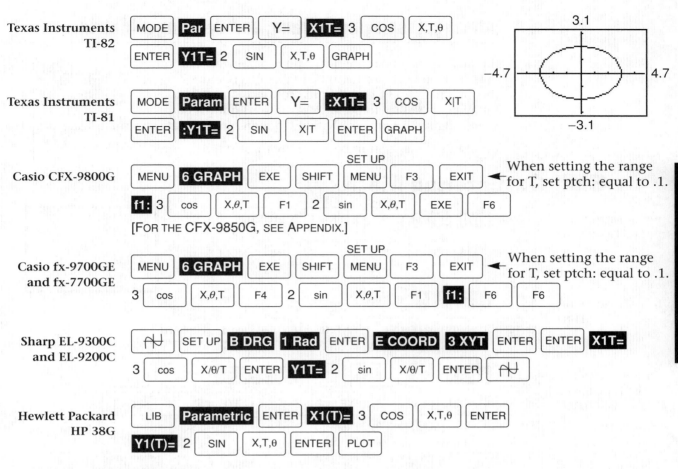

Texas Instruments
TI-82

| MODE | **Par** | ENTER | Y= | **X1T=** | 3 | COS | X,T,θ |

| ENTER | **Y1T=** | 2 | SIN | X,T,θ | GRAPH |

Texas Instruments
TI-81

| MODE | **Param** | ENTER | Y= | **:X1T=** | 3 | COS | X|T |

| ENTER | **:Y1T=** | 2 | SIN | X|T | ENTER | GRAPH |

Casio CFX-9800G

SET UP

| MENU | **6 GRAPH** | EXE | SHIFT | MENU | F3 | EXIT |

f1: | 3 | cos | X,θ,T | F1 | 2 | sin | X,θ,T | EXE | F6 |

When setting the range
for T, set ptch: equal to .1.

[FOR THE CFX-9850G, SEE APPENDIX.]

Casio fx-9700GE
and fx-7700GE

SET UP

| MENU | **6 GRAPH** | EXE | SHIFT | MENU | F3 | EXIT |

| 3 | cos | X,θ,T | F4 | 2 | sin | X,θ,T | F1 | **f1:** | F6 | F6 |

When setting the range
for T, set ptch: equal to .1.

Sharp EL-9300C
and EL-9200C

| ⋔ | SET UP | **B DRG** | **1 Rad** | ENTER | **E COORD** | **3 XYT** | ENTER | ENTER | **X1T=** |

| 3 | cos | X/θ/T | ENTER | **Y1T=** | 2 | sin | X/θ/T | ENTER | ⋔ |

Hewlett Packard
HP 38G

| LIB | **Parametric** | ENTER | **X1(T)=** | 3 | COS | X,T,θ | ENTER |

| **Y1(T)=** | 2 | SIN | X,T,θ | ENTER | PLOT |

INVESTIGATIONS

1. The equation $ax^2 + by^2 = 36$ represents a family of ellipses.

   a. Try experimenting with values of a and b that are different.

   b. Try experimenting with values of a and b that are the same.

   c. Investigate what happens when you graph an ellipse whose equation is $4x^2 + 25y^2 = 100$ and then graph the ellipse for which 4 and 25 are switched.

2. The equation $x^2 - y^2 = c^2$ represents a family of hyperbolas. When the single equation is written as two equations, you will have the pair shown here. The number c can be any positive number large or small.

$$\begin{cases} y = \sqrt{x^2 - c^2} \\ y = -\sqrt{x^2 - c^2} \end{cases}$$

   a. Try investigating the effect of choosing a large value of c, such as 5 or 10.

   b. Try investigating the effect of choosing a small value of c, such as 0.5 or 0.1.

   c. What do you think will happen if c is allowed to become very small and approach 0? [The hyperbola will become a pair of lines with slope 1 and −1 and intersect at the origin.]

3. Every ellipse is a bounded curve. Another way to say this fact about an ellipse is that every ellipse sits inside a rectangle. The pair of parametric equations shown here will model an ellipse whose center is the point with coordinates (c, d).

$$\begin{cases} x(t) = a \cos t + c \\ y(t) = b \sin t + d \\ 0 \le t \le 2\pi \end{cases}$$

   Experiment with different values of a and b to make an ellipse with center at the origin and that is inside the rectangle bounded by $-6 \le x \le 6$ and $-4 \le y \le 4$.

Entering, Displaying, Editing, and Deleting a Matrix

In many situations, numerical data is organized in a table. The numerical data in this table fills two rows (rows 1 and 2) and two columns (columns 1 and 2). It may be necessary to perform operations on this data. Consequently, it may be advisable to represent the data in a matrix. A *matrix* is a rectangular array of numbers. Follow these steps to enter matrix $A = \begin{bmatrix} 3 & 4 \\ 1 & 5 \end{bmatrix}$.

	col 1	col 2
row 1	3	4
row 2	1	5

Skill 1: Entering Matrix A

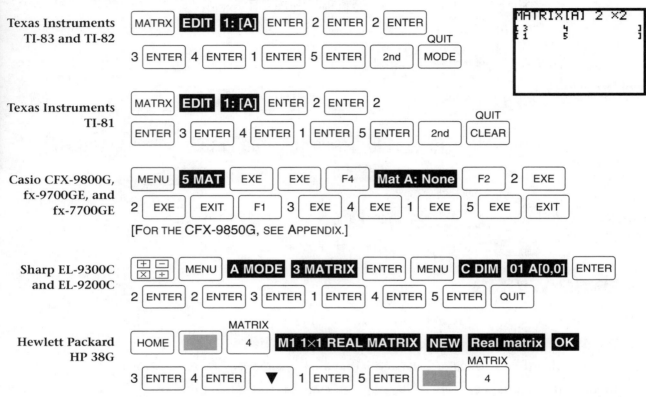

To display matrix A at some time after you have created it, follow these steps.

Skill 2: Displaying Matrix A

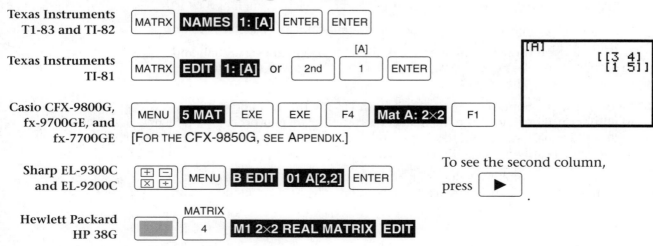

To see the second column, press ▶ .

To edit matrix A to read, for example, $\begin{bmatrix} 7 & -2 \\ 7 & 1 \end{bmatrix}$, follow these steps.

Skill 3: Editing Matrix *A*

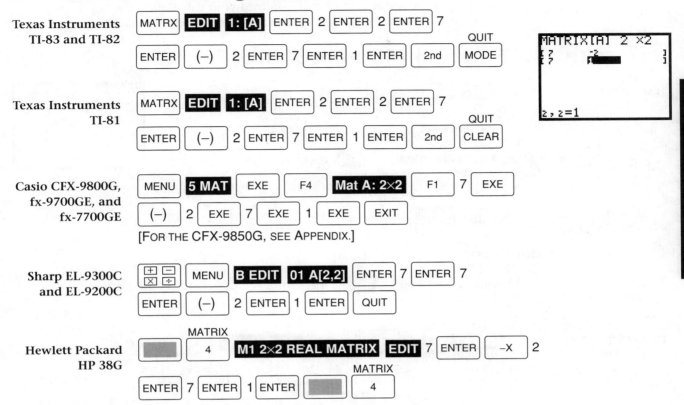

Texas Instruments TI-83 and TI-82
`MATRX` `EDIT` `1: [A]` `ENTER` 2 `ENTER` 2 `ENTER` 7
`ENTER` `(−)` 2 `ENTER` 7 `ENTER` 1 `ENTER` `2nd` `MODE` (QUIT)

Texas Instruments TI-81
`MATRX` `EDIT` `1: [A]` `ENTER` 2 `ENTER` 2 `ENTER` 7
`ENTER` `(−)` 2 `ENTER` 7 `ENTER` 1 `ENTER` `2nd` `CLEAR` (QUIT)

Casio CFX-9800G, fx-9700GE, and fx-7700GE
`MENU` `5 MAT` `EXE` `F4` `Mat A: 2×2` `F1` 7 `EXE`
`(−)` 2 `EXE` 7 `EXE` 1 `EXE` `EXIT`
[FOR THE CFX-9850G, SEE APPENDIX.]

Sharp EL-9300C and EL-9200C
`[+ − × ÷]` `MENU` `B EDIT` `01 A[2,2]` `ENTER` 7 `ENTER` 7
`ENTER` `(−)` 2 `ENTER` 1 `ENTER` `QUIT`

Hewlett Packard HP 38G
`[ ]` `4` (MATRIX) `M1 2×2 REAL MATRIX` `EDIT` 7 `ENTER` `−X` 2
`ENTER` 7 `ENTER` 1 `ENTER` `[ ]` `4` (MATRIX)

You may need to delete matrix A. Here is how to do it. Use the down arrow to highlight the indicated menu choice, such as `2: Delete...`, and press `ENTER` to select it.

Skill 4: Deleting Matrix *A*

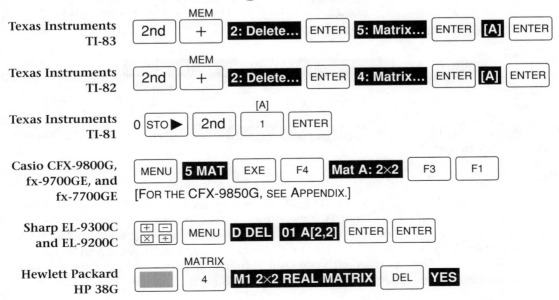

Texas Instruments TI-83
`2nd` `+` (MEM) `2: Delete...` `ENTER` `5: Matrix...` `ENTER` `[A]` `ENTER`

Texas Instruments TI-82
`2nd` `+` (MEM) `2: Delete...` `ENTER` `4: Matrix...` `ENTER` `[A]` `ENTER`

Texas Instruments TI-81
0 `STO▶` `2nd` `1` [A] `ENTER`

Casio CFX-9800G, fx-9700GE, and fx-7700GE
`MENU` `5 MAT` `EXE` `F4` `Mat A: 2×2` `F3` `F1`
[FOR THE CFX-9850G, SEE APPENDIX.]

Sharp EL-9300C and EL-9200C
`[+ − × ÷]` `MENU` `D DEL` `01 A[2,2]` `ENTER` `ENTER`

Hewlett Packard HP 38G
`[ ]` `4` (MATRIX) `M1 2×2 REAL MATRIX` `DEL` `YES`

GRAPHICS CALCULATOR SKILLS

Matrix Operations and Inverses

Suppose you have $A = \begin{bmatrix} 3 & 4 \\ 1 & 5 \end{bmatrix}$ and $B = \begin{bmatrix} -2 & 1 \\ 0 & -10 \end{bmatrix}$. To find the sum, difference, or product of A and B, first enter them into your calculator. The key sequences that follow illustrate how to find a sum of two compatible matrices.

Skill 1: Entering Matrices A and B

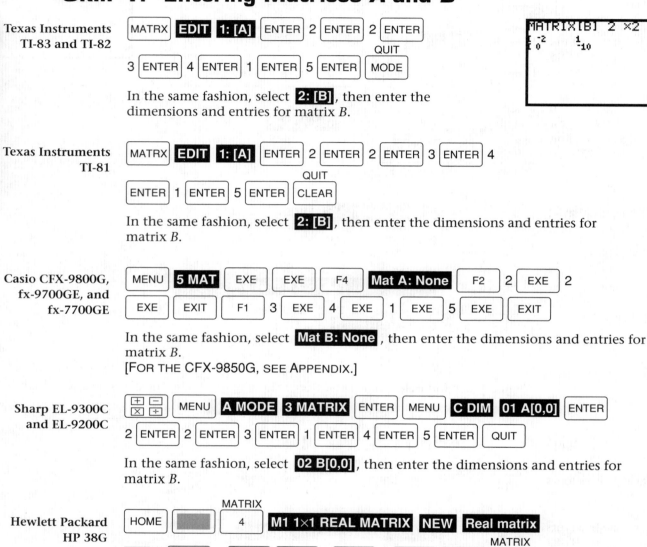

Texas Instruments TI-83 and TI-82

MATRX | EDIT | 1: [A] | ENTER | 2 | ENTER | 2 | ENTER

3 | ENTER | 4 | ENTER | 1 | ENTER | 5 | ENTER | MODE (QUIT)

In the same fashion, select **2: [B]**, then enter the dimensions and entries for matrix B.

```
MATRIX[B] 2 ×2
[ -2    1      ]
[ 0    -10     ]
```

Texas Instruments TI-81

MATRX | EDIT | 1: [A] | ENTER | 2 | ENTER | 2 | ENTER | 3 | ENTER | 4

ENTER | 1 | ENTER | 5 | ENTER | CLEAR (QUIT)

In the same fashion, select **2: [B]**, then enter the dimensions and entries for matrix B.

Casio CFX-9800G, fx-9700GE, and fx-7700GE

MENU | 5 MAT | EXE | EXE | F4 | Mat A: None | F2 | 2 | EXE | 2

EXE | EXIT | F1 | 3 | EXE | 4 | EXE | 1 | EXE | 5 | EXE | EXIT

In the same fashion, select **Mat B: None**, then enter the dimensions and entries for matrix B.
[FOR THE CFX-9850G, SEE APPENDIX.]

Sharp EL-9300C and EL-9200C

[+ −] [× ÷] | MENU | A MODE | 3 MATRIX | ENTER | MENU | C DIM | 01 A[0,0] | ENTER

2 | ENTER | 2 | ENTER | 3 | ENTER | 1 | ENTER | 4 | ENTER | 5 | ENTER | QUIT

In the same fashion, select **02 B[0,0]**, then enter the dimensions and entries for matrix B.

Hewlett Packard HP 38G

HOME | [MATRIX] 4 | M1 1×1 REAL MATRIX | NEW | Real matrix

OK | 3 | ENTER | 4 | ENTER | ▼ | 1 | ENTER | 5 | ENTER | [MATRIX] 4

In the same fashion, select **M2 1×1 REAL MATRIX**, then enter the dimensions and entries for matrix B.

You can perform operations on matrices if they are compatible. If matrix A has dimensions $m \times n$ and matrix B has dimensions $m \times n$, you can form the sum A + B and the difference A − B. If matrix A has dimensions $m \times n$ and matrix B has dimensions $n \times p$ you can form the product AB. Since matrices A and B given here are square matrices with the same dimensions, you can add, subtract, and multiply them. The procedure for each operation is the same. To subtract or multiply, simply replace [+] with [−] or [×].

Skill 2: Finding the Sum of *A* and *B*

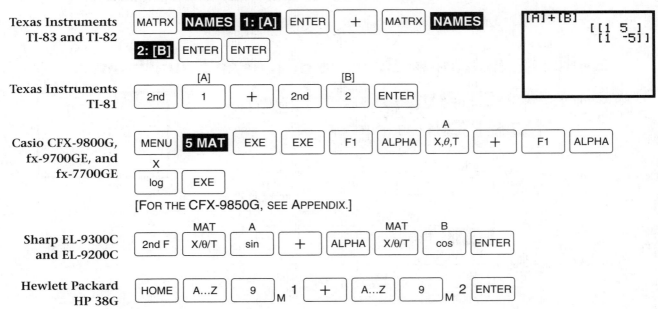

Texas Instruments TI-83 and TI-82: MATRX NAMES 1: [A] ENTER + MATRX NAMES 2: [B] ENTER ENTER

Texas Instruments TI-81: 2nd 1 [A] + 2nd 2 [B] ENTER

Casio CFX-9800G, fx-9700GE, and fx-7700GE: MENU 5 MAT EXE EXE F1 ALPHA X,θ,T [A] + F1 ALPHA X [log] EXE
[FOR THE CFX-9850G, SEE APPENDIX.]

Sharp EL-9300C and EL-9200C: 2nd F X/θ/T [MAT] sin [A] + ALPHA X/θ/T [MAT] cos [B] ENTER

Hewlett Packard HP 38G: HOME A...Z 9 M 1 + A...Z 9 M 2 ENTER

[A]+[B]
[[1 5]
 [1 -5]]

You can use your graphics calculator along with the inverse of a matrix and matrix multiplication to solve a system of linear equations if that system has as many unknowns as it has equations.

$$\begin{cases} 3x + 4y = 12 \\ x + 5y = -10 \end{cases} \longrightarrow \begin{bmatrix} 3 & 4 \\ 1 & 5 \end{bmatrix} \begin{bmatrix} x \\ y \end{bmatrix} = \begin{bmatrix} 12 \\ -10 \end{bmatrix}$$

The solution of the matrix equation is $\begin{bmatrix} x \\ y \end{bmatrix} = \begin{bmatrix} 3 & 4 \\ 1 & 5 \end{bmatrix}^{-1} \begin{bmatrix} 12 \\ -10 \end{bmatrix}$. First, enter $\begin{bmatrix} 3 & 4 \\ 1 & 5 \end{bmatrix}$ as matrix *A*. Enter $\begin{bmatrix} 12 \\ -10 \end{bmatrix}$ as matrix *B*. Then the product $A^{-1}B$ will give $\begin{bmatrix} x \\ y \end{bmatrix}$. Use the appropriate key sequence that follows to find the inverse of $A = \begin{bmatrix} 3 & 4 \\ 1 & 5 \end{bmatrix}$. Be sure that matrix *A* is entered into the calculator and that you are in matrix mode.

Skill 3: Finding A^{-1}

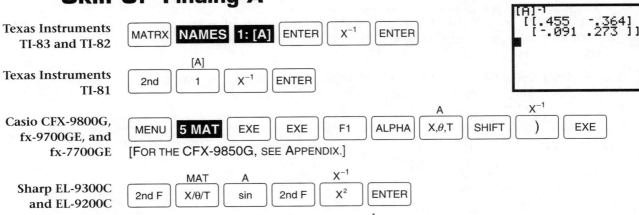

Texas Instruments TI-83 and TI-82: MATRX NAMES 1: [A] ENTER X^{-1} ENTER

Texas Instruments TI-81: 2nd 1 [A] X^{-1} ENTER

Casio CFX-9800G, fx-9700GE, and fx-7700GE: MENU 5 MAT EXE EXE F1 ALPHA X,θ,T [A] SHIFT) X^{-1} EXE
[FOR THE CFX-9850G, SEE APPENDIX.]

Sharp EL-9300C and EL-9200C: 2nd F X/θ/T [MAT] sin [A] 2nd F X^2 X^{-1} ENTER

Hewlett Packard HP 38G: HOME A...Z 9 M 1 [ ] X,T,θ X^{-1} ENTER

[A]⁻¹
[[.455 -.364]
 [-.091 .273]]

GRAPHICS CALCULATOR SKILLS

To solve the system preceding Skill 3 and shown here, combine the operation of finding an inverse of a matrix and the operation of matrix multiplication.

$$\begin{bmatrix} 3 & 4 \\ 1 & 5 \end{bmatrix} \begin{bmatrix} x \\ y \end{bmatrix} = \begin{bmatrix} 12 \\ -10 \end{bmatrix}$$

Skill 4: Solving a System of Linear Equations

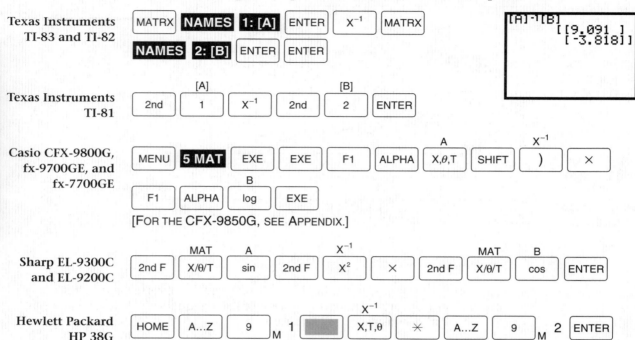

INVESTIGATIONS

1. a. Try investigating some of these systems of equations.

$$\begin{cases} 3x + 4y = -12 \\ x + 5y = 10 \end{cases} \qquad \begin{cases} 3x + 4y = 12 \\ 3x + 4y = -10 \end{cases} \qquad \begin{cases} 3x + 4y = 12 \\ 6x + 8y = 24 \end{cases}$$

   b. Try exploring the graphs of the equations in the systems you chose to investigate.

   c. The algebraic and the geometric results indicate that a pair of linear equations with two unknowns has a unique solution, no solution at all, or infinitely many solutions. Explore other systems of linear equations to confirm this statement.

2. Try exploring these related systems. Look for ways you can reduce the work needed to solve all of them. Try your methods on other related systems.

$$\begin{cases} 3x + 4y = 12 \\ x + 5y = -10 \end{cases} \qquad \begin{cases} 3x + 4y = 3 \\ x + 5y = -8 \end{cases} \qquad \begin{cases} 3x + 4y = 0 \\ x + 5y = 4 \end{cases}$$

[9.09, -3.82]   [4.27, -2.45]   [-1.45, 1.09]

GRAPHICS CALCULATOR SKILLS

Introduction to One- and Two-Variable Data

Statistical measures of numerical data are fast becoming an important part of everyone's life. An example of one-variable data is the set of weights of cans filled with a product and taken from the manufacturing line for sampling. In this instance, an analyst may want to know how close to a predetermined weight the final product is likely to be.

Graphics calculators have the capability of providing measures of central tendency, such as mean or average weight, and measures of dispersion, such as range and standard deviation. First, however, you will need to see how to enter a given set of data.

The following keystrokes illustrate the steps needed to enter this data set.

$$\{6, 9, 11.4, 8, 9\}$$

Note: Before you enter this data, you may need to clear previously entered data from the calculator. Skill 4 shows how to do this. Use the down arrow to highlight the indicated menu choice, such as **1: Edit...**, and press ENTER to select it.

Skill 1: Entering One-Variable Data

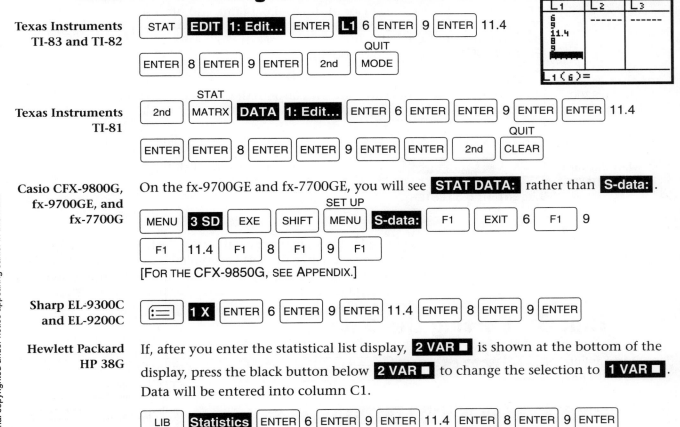

Texas Instruments TI-83 and TI-82

STAT **EDIT** **1: Edit...** ENTER **L1** 6 ENTER 9 ENTER 11.4

ENTER 8 ENTER 9 ENTER 2nd MODE *(QUIT)*

Texas Instruments TI-81

2nd MATRX *(STAT)* **DATA** **1: Edit...** ENTER 6 ENTER ENTER 9 ENTER ENTER 11.4

ENTER ENTER 8 ENTER ENTER 9 ENTER ENTER 2nd CLEAR *(QUIT)*

Casio CFX-9800G, fx-9700GE, and fx-7700G

On the fx-9700GE and fx-7700GE, you will see **STAT DATA:** rather than **S-data:**.

MENU **3 SD** EXE SHIFT MENU *(SET UP)* **S-data:** F1 EXIT 6 F1 9

F1 11.4 F1 8 F1 9 F1

[FOR THE CFX-9850G, SEE APPENDIX.]

Sharp EL-9300C and EL-9200C

▤ **1 X** ENTER 6 ENTER 9 ENTER 11.4 ENTER 8 ENTER 9 ENTER

Hewlett Packard HP 38G

If, after you enter the statistical list display, **2 VAR ▪** is shown at the bottom of the display, press the black button below **2 VAR ▪** to change the selection to **1 VAR ▪**. Data will be entered into column C1.

LIB **Statistics** ENTER 6 ENTER 9 ENTER 11.4 ENTER 8 ENTER 9 ENTER

You may need to edit a data set for some reason.

- A data entry error was made, perhaps an omission, a typing error, or a repetition.
- You have completed the study of one data set and want to study a related data set.

Suppose, for example, that you enter {6, 9, 11.4, 8, 9}, need to change one 9 to 9.5, and insert 6.8. One strategy for editing the data set is to place the cursor in a data line that contains one of the nines, perhaps the second 9. Change 9 to 9.5. Enter or insert 6.8 into a blank space in the list.

The implementation of this strategy is shown here.

Skill 2: Editing a Data Set

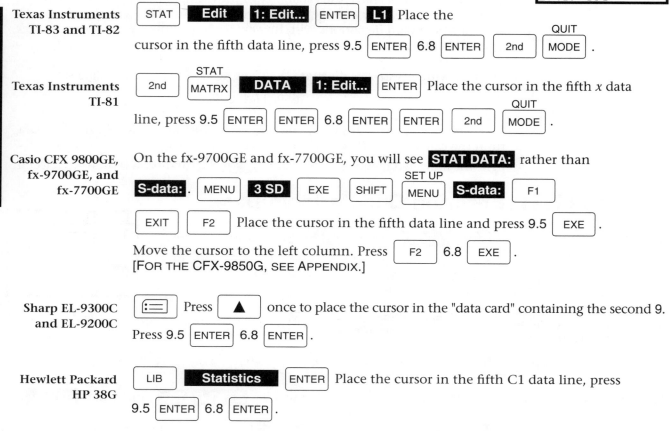

Texas Instruments TI-83 and TI-82   [STAT] **Edit** **1: Edit...** [ENTER] **L1** Place the cursor in the fifth data line, press 9.5 [ENTER] 6.8 [ENTER] [2nd] [MODE] (QUIT).

Texas Instruments TI-81   [2nd] [MATRX] (STAT) **DATA** **1: Edit...** [ENTER] Place the cursor in the fifth *x* data line, press 9.5 [ENTER] [ENTER] 6.8 [ENTER] [ENTER] [2nd] [MODE] (QUIT).

Casio CFX 9800GE, fx-9700GE, and fx-7700GE   On the fx-9700GE and fx-7700GE, you will see **STAT DATA:** rather than **S-data:**. [MENU] **3 SD** [EXE] [SHIFT] [MENU] (SET UP) **S-data:** [F1] [EXIT] [F2] Place the cursor in the fifth data line and press 9.5 [EXE]. Move the cursor to the left column. Press [F2] 6.8 [EXE].
[FOR THE CFX-9850G, SEE APPENDIX.]

Sharp EL-9300C and EL-9200C   [≔] Press [▲] once to place the cursor in the "data card" containing the second 9. Press 9.5 [ENTER] 6.8 [ENTER].

Hewlett Packard HP 38G   [LIB] **Statistics** [ENTER] Place the cursor in the fifth C1 data line, press 9.5 [ENTER] 6.8 [ENTER].

Once the data is entered, some calculators will sort it. Such a sort can help you find the median of a data set. Once the sort is completed, follow the key sequence that displays the data. The display shown here illustrates the edited data from Skill 2 sorted in ascending order. [FOR THE CFX-9850G, SEE APPENDIX.]

Skill 3: Sorting the Data

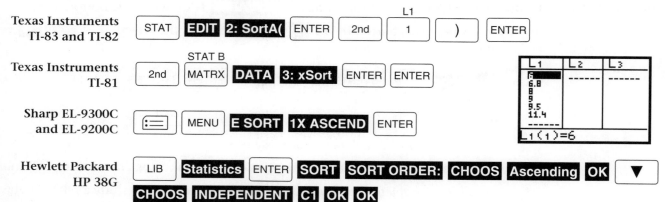

Texas Instruments TI-83 and TI-82   [STAT] **EDIT** **2: SortA(** [ENTER] [2nd] **1** (L1) **)** [ENTER]

Texas Instruments TI-81   [2nd] [MATRX] (STAT B) **DATA** **3: xSort** [ENTER] [ENTER]

Sharp EL-9300C and EL-9200C   [≔] [MENU] **E SORT** **1X ASCEND** [ENTER]

Hewlett Packard HP 38G   [LIB] **Statistics** [ENTER] **SORT** **SORT ORDER:** **CHOOS** **Ascending** **OK** [▼] **CHOOS** **INDEPENDENT** **C1** **OK** **OK**

Skill 4: Clearing Previously Entered Data

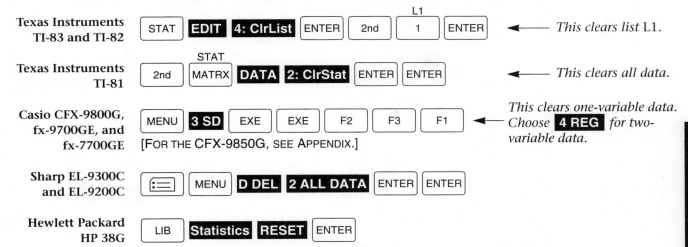

Texas Instruments TI-83 and TI-82

| STAT | **EDIT** | **4: ClrList** | ENTER | 2nd | L1 1 | ENTER | ← *This clears list L1.*

Texas Instruments TI-81

| 2nd | MATRX STAT | **DATA** | **2: ClrStat** | ENTER | ENTER | ← *This clears all data.*

Casio CFX-9800G, fx-9700GE, and fx-7700GE

| MENU | **3 SD** | EXE | EXE | F2 | F3 | F1 | ← *This clears one-variable data. Choose* **4 REG** *for two-variable data.*

[FOR THE CFX-9850G, SEE APPENDIX.]

Sharp EL-9300C and EL-9200C

| :≡ | MENU | **D DEL** | **2 ALL DATA** | ENTER | ENTER |

Hewlett Packard HP 38G

| LIB | **Statistics** | **RESET** | ENTER |

In many statistical problems, you need to deal with data that involves two variables. The following key sequences show how to enter this two-variable set of data: {(0, 3.5), (1, 3.9), (2, 4.2), (3, 5.8), (4, 6.0), (5, 6.5)}.

Skill 5: Entering Two-Variable Data

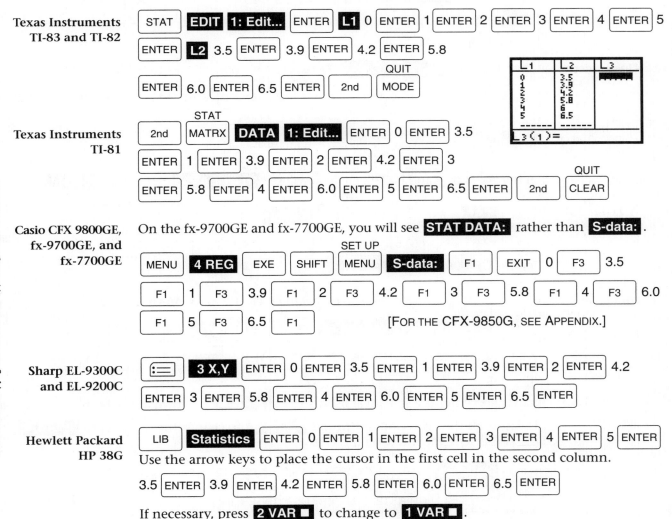

Texas Instruments TI-83 and TI-82

| STAT | **EDIT** | **1: Edit...** | ENTER | **L1** 0 ENTER 1 ENTER 2 ENTER 3 ENTER 4 ENTER 5

ENTER | **L2** 3.5 ENTER 3.9 ENTER 4.2 ENTER 5.8

ENTER 6.0 ENTER 6.5 ENTER | 2nd | QUIT MODE

Texas Instruments TI-81

| 2nd | MATRX STAT | **DATA** | **1: Edit...** | ENTER 0 ENTER 3.5

ENTER 1 ENTER 3.9 ENTER 2 ENTER 4.2 ENTER 3

ENTER 5.8 ENTER 4 ENTER 6.0 ENTER 5 ENTER 6.5 ENTER | 2nd | QUIT CLEAR

Casio CFX 9800GE, fx-9700GE, and fx-7700GE

On the fx-9700GE and fx-7700GE, you will see **STAT DATA:** rather than **S-data:**.

| MENU | **4 REG** | EXE | SHIFT | SET UP MENU | **S-data:** | F1 | EXIT 0 F3 3.5

F1 1 F3 3.9 F1 2 F3 4.2 F1 3 F3 5.8 F1 4 F3 6.0

F1 5 F3 6.5 F1   [FOR THE CFX-9850G, SEE APPENDIX.]

Sharp EL-9300C and EL-9200C

| :≡ | **3 X,Y** | ENTER 0 ENTER 3.5 ENTER 1 ENTER 3.9 ENTER 2 ENTER 4.2

ENTER 3 ENTER 5.8 ENTER 4 ENTER 6.0 ENTER 5 ENTER 6.5 ENTER

Hewlett Packard HP 38G

| LIB | **Statistics** | ENTER 0 ENTER 1 ENTER 2 ENTER 3 ENTER 4 ENTER 5 ENTER

Use the arrow keys to place the cursor in the first cell in the second column.

3.5 ENTER 3.9 ENTER 4.2 ENTER 5.8 ENTER 6.0 ENTER 6.5 ENTER

If necessary, press **2 VAR ■** to change to **1 VAR ■**.

GRAPHICS CALCULATOR SKILLS

Measures of Central Tendency and Dispersion

A descriptive statistical list contains information that indicates how a set of data clusters in some way and how that set of data tends to spread out or disperse. This table provides a short list of some statistical terms used in the measurement of clustering and dispersion.

Central Tendency	Dispersion
mean	range
median	variance
mode	standard deviation

Skill 1 illustrates the steps needed to enter {12.1, 13.2, 12.8, 12.5, 12.6, 12.4, 12.7, 12.1}. Skill 2 will show how to calculate measures of central tendency. Finally, Skill 3 will take care of the work needed to measure dispersion. Use the down arrow to highlight the indicated menu choice, such as **1: Edit...**, and press ENTER to select it.

Skill 1: Entering the Data

Texas Instruments TI-83 and TI-82

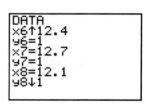

STAT **EDIT** **1: Edit...** ENTER **L1** 12.1 ENTER 13.2

ENTER 12.8 ENTER 12.5 ENTER 12.6 ENTER 12.4

ENTER 12.7 ENTER 12.1 ENTER 2nd MODE (QUIT)

Texas Instruments TI-81

2nd MATRX (STAT) **DATA** **1: Edit...** ENTER 12.1 ENTER ENTER 13.2 ENTER ENTER

12.8 ENTER ENTER 12.5 ENTER ENTER 12.6 ENTER ENTER 12.4 ENTER ENTER 12.7

ENTER ENTER 12.1 ENTER ENTER 2nd CLEAR (QUIT)

Casio CFX-9800G, fx-9700GE, and fx-7700GE

On the fx-9700GE and fx-7700GE, you will see **STAT DATA:** rather than **S-data:** .

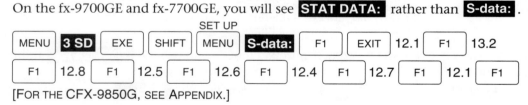

MENU **3 SD** EXE SHIFT MENU (SET UP) **S-data:** F1 EXIT 12.1 F1 13.2

F1 12.8 F1 12.5 F1 12.6 F1 12.4 F1 12.7 F1 12.1 F1

[FOR THE CFX-9850G, SEE APPENDIX.]

Sharp EL-9300C and EL-9200C

Note: According to these keystrokes, begin by clearing any previous data.

:≡ MENU **D DEL** **2 ALL DATA** ENTER ENTER **1X** ENTER 12.1 ENTER 13.2

ENTER 12.8 ENTER 12.5 ENTER 12.6 ENTER 12.4 ENTER 12.7 ENTER 12.1 ENTER

Hewlett Packard HP 38G

If, after you enter the statistical list display, **2 VAR ■** is shown at the bottom of the display, press the black button below 2 **2 VAR ■** to change the selection to **1 VAR ■**. Data will be entered into Column C1.

LIB **Statistics** ENTER 12.1 ENTER 13.2 ENTER 12.8 ENTER 12.5 ENTER

12.6 ENTER 12.4 ENTER 12.7 ENTER 12.1 ENTER

The *mean* of a set of numerical data is defined as the sum of all the data values divided by the number of data values. The *median* of a data set having an odd number of data values is that data value in the middle of the data set when it is arranged in ascending order or descending order. If the number of data values is even, the *median* is the average of the two data values in the middle.

The keystrokes shown in Skill 2 illustrate the calculation of the mean. Notice that some calculators provide a fairly full statistical report. On other calculators, you press particular keys to calculate a particular measure. With the data entered as described in Skill 1, follow the appropriate key sequence. Use the down arrow to highlight the indicated menu choice.

GRAPHICS CALCULATOR SKILLS

Skill 2: Calculating the Mean and Median

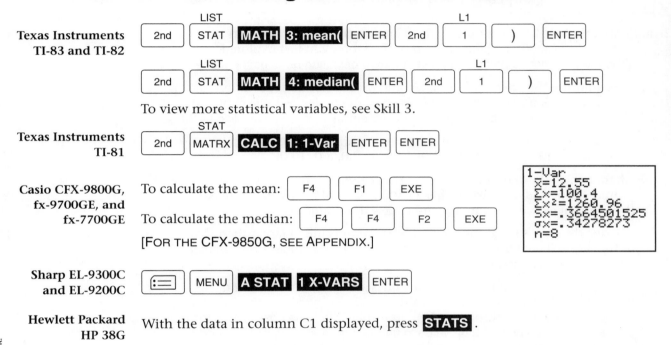

Texas Instruments TI-83 and TI-82

LIST
2nd · STAT · **MATH** · **3: mean(** · ENTER · 2nd · 1 ·) · ENTER   (L1)

LIST
2nd · STAT · **MATH** · **4: median(** · ENTER · 2nd · 1 ·) · ENTER   (L1)

To view more statistical variables, see Skill 3.

Texas Instruments TI-81

STAT
2nd · MATRX · **CALC** · **1: 1-Var** · ENTER · ENTER

Casio CFX-9800G, fx-9700GE, and fx-7700GE

To calculate the mean: F4 · F1 · EXE

To calculate the median: F4 · F4 · F2 · EXE

[FOR THE CFX-9850G, SEE APPENDIX.]

```
1-Var
x̄=12.55
Σx=100.4
Σx²=1260.96
Sx=.3664501525
σx=.34278273
n=8
```

Sharp EL-9300C and EL-9200C

[≔] · MENU · **A STAT** · **1 X-VARS** · ENTER

Hewlett Packard HP 38G

With the data in column C1 displayed, press **STATS** .

Standard deviation is a measure of how data spreads out from the mean of the data set. The term, however, can be used in two contexts. It can refer to a sample, which usually is a small set. The calculator notation could be Sx, (Xsn – 1), or SSDEV. The term can also refer to a population, frequently a set much larger than a sample. In the case of a population, the standard deviation is most likely an estimate. The calculator notation for population standard deviation could be σx, (Xσn), or PSDEV.

Skill 3: Calculating the Standard Deviation

To calculate standard deviation on a TI-81, Sharp, or Hewlett Packard graphics calculator, see Skill 2.

Texas Instruments TI-83 and TI-82

Enter the data as described, press STAT · **CALC** · **1:** · ENTER · ENTER .

Casio CFX-9800G, fx-9700GE, and fx-7700GE

Enter the data as described, press EXIT · F4 · F3 · EXE for the sample standard deviation.

Enter the data as described, press EXIT · F4 · F2 · EXE for the population standard deviation.
[FOR THE CFX-9850G, SEE APPENDIX.]

INVESTIGATIONS

1. Try experimenting with some of the data sets that follow to see how their means and standard deviations compare.

Table A	Table B	Table C	Table D
x	x	x	x
12.5	13.1	13.4	11.3
12.5	12.9	13.1	11.7
12.5	12.7	12.8	12.1
12.5	12.5	12.5	12.5
12.5	12.3	12.8	12.1
12.5	12.1	13.1	11.7
12.5	12.9	13.4	11.3

2. **a.** Try exploring the effect on the mean and on the standard deviation of adding 3 to each data value in each of the tables you just explored. [The mean is increased by 3 and the standard deviation remains the same.]

   b. Try exploring the effect on the mean and on the standard deviation of multiplying each data value by 3 in each of the tables you just explored. [The mean is multiplied by 3 and the standard deviation is also multiplied by 3.]

3. Try modifying this data set so that it has the required characteristic.
   {0.5, 1.3, 2.4, 3.1, 3.9, 4.5, 6.6, 7.0, 8.0, 10.0}

   a. Change the data so that the mean of the revised data set is negative. Think of different ways to do this. (Try thinking of how you can make the mean negative by changing only a few data values.)

   b. Change the data so that the standard deviation is much larger than that of the original data set. (Think of making small data values even smaller and large data values even larger.)

   c. Change the data so that all the data values are whole numbers and the new data set has a whole number as the mean. Think about the fact that the sum of the data values must be a multiple of 8. [One possibility is to increase each nonintegral data value to the next greater integer. Then add 5 to any one of the new data values. The data sum will be 56 and the mean will be 7.]

4. Try creating some data sets of your own that have the following characteristics. Use the calculator to confirm your conclusions and edit your data to revise your data set as you go.

   a. a data set that has seven data values all different and whose mean is between 5 and 10

   b. a data set that has seven data values all different and whose standard deviation is close to 0

   c. a data set that has seven data values all different and whose mean is positive and less than 3

   d. a data set that has seven data values all different and whose standard deviation is larger than 10

5. Try using different data sets and the calculator to investigate this statement. The average of the numbers in a finite set of different numbers must be somewhere between the smallest and the largest of them.

GRAPHICS CALCULATOR SKILLS

Histograms

In many real-life sampling and probability applications, practitioners find that some data values occur many times in the statistical sample or sample space. Indeed, the statistical concepts of *mode* and *frequency* were born of this phenomenon. For example, a quality control analyst may find that, in a sample of fifty cans of soup that are to weigh 12 oz each, exactly twenty-four cans weigh 11.999 oz and exactly ten cans weigh 12.001 oz. Clearly the mode, the most frequent data value, must be 11.999 oz. Consider this relatively simple data set.
{4.2, 3.1, 5.8, 5.1, 4.2, 2.5, 6.8, 7.2, 5.1, 3.1, 5.8, 4.2}

You can display the data in a *histogram*, a statistical chart for one-variable data that indicates the frequency of each data value. The instructions for creating a histogram on the Texas Instruments, Sharp, and Hewlett Packard calculators are given here. The instructions for Casio calculators will be given separately.

Data Value	Frequency
2.5	1
3.1	2
4.2	3
5.1	2
5.8	2
6.8	1
7.2	1

Skill 1: Entering the Data

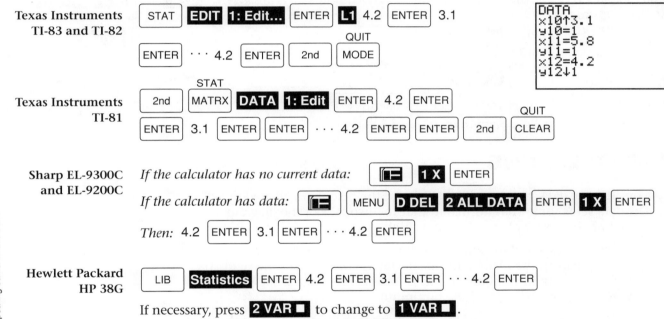

Texas Instruments TI-83 and TI-82

STAT EDIT 1: Edit... ENTER L1 4.2 ENTER 3.1

ENTER · · · 4.2 ENTER 2nd MODE (QUIT)

DATA
×10↑3.1
y10=1
×11=5.8
y11=1
×12=4.2
y12↓1

Texas Instruments TI-81

2nd MATRX (STAT) DATA 1: Edit ENTER 4.2 ENTER

ENTER 3.1 ENTER ENTER · · · 4.2 ENTER ENTER 2nd CLEAR (QUIT)

Sharp EL-9300C and EL-9200C

If the calculator has no current data: ▣ 1 X ENTER

If the calculator has data: ▣ MENU D DEL 2 ALL DATA ENTER 1 X ENTER

Then: 4.2 ENTER 3.1 ENTER · · · 4.2 ENTER

Hewlett Packard HP 38G

LIB Statistics ENTER 4.2 ENTER 3.1 ENTER · · · 4.2 ENTER

If necessary, press 2 VAR ■ to change to 1 VAR ■ .

Now that the data is entered, you are ready to create the histogram.

- Clear any previously entered functions from the function list.

- If you are using a Texas Instruments or Sharp calculator, use the viewing window shown. This step is not necessary for Hewlett Packard HP 38G users.

Skill 2: Creating the Histogram

Texas Instruments TI-83 and TI-82

2nd Y= (STAT PLOT) 1: Plot 1... ENTER On ENTER ▊▎

ENTER Xlist: L1 ENTER Freq: 1 ENTER GRAPH

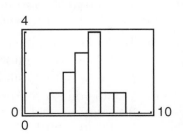

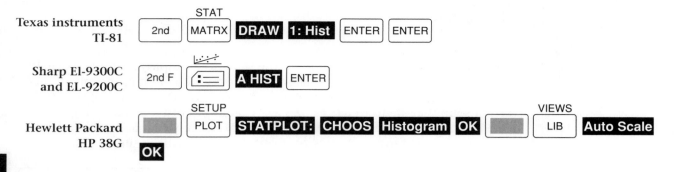

Texas instruments TI-81

Sharp El-9300C and EL-9200C

Hewlett Packard HP 38G

The keystrokes that follow illustrate how to create a histogram on the Casio graphics calculators. First, identify intervals for grouping the data. For example, the intervals shown here would be quite suitable.

| 2-3 | 3-4 | 4-5 | 5-6 | 6-7 | 7-8 |

The number of intervals, 6, will be entered as the number of bars in the bar graph. For this set of data, you must use the viewing window settings $2 \leq x \leq 8$ and $0 \leq y \leq 4$. Be sure to specify these settings before you enter the number of bars.

Casio CFX-9800G, fx-9700GE, and fx-7700GE

On the fx-9700GE and fx-7700GE, the labels are **STAT DATA:** and **STAT GRAPH:**.

| MENU | **3 SD** | EXE | SHIFT | 3 | F2 | EXE | SHIFT | MENU | **S-data:** |

CLR

SET UP

| F1 | **S-Graph** | F1 | EXIT | SHIFT | • | 6 | EXE | 4.2 | F1 | 3.1 | F1 | 5.8 |

Defm

| F1 | ··· 4.2 | F1 | GRAPH | EXE |

[FOR THE CFX-9850G, SEE APPENDIX.]

INVESTIGATIONS

1. Consider the data set entered in Skill 1 and the histogram created in Skill 2. The data set is somewhat normal in that the two tails of the diagram are low while the center of the diagram is high.

   a. A data set is skewed toward the minimum if the peak is toward the left and the bars in the histogram get shorter as you move to the right. Try experimenting with modifications in the data set so that the set and histogram are skewed toward the minimum.

   b. A data set is skewed toward the maximum if the peak is toward the right and the bars in the histogram get longer as you move to the right. Try experimenting with modifications in the data set so that the set and histogram are skewed toward the maximum.

   c. Consider different ways to modify the data set so that the histogram has a dip or valley in the middle instead of a peak.

2. A distribution is uniform, or nearly so, if all the frequencies of the data values are the same or about the same.

   a. Try looking for different ways to modify the given data set so that it is nearly uniform, but yet has a slight peak in the center.

   b. Consider the experiment of flipping a coin or rolling a number cube. Would you expect the distribution resulting from the experiment to be uniform? Consider what this means in terms of the phrase "equally likely" in probability.

Box-and-Whisker Plots

Listed here are some commonly used statistical measures that can be applied to one-variable data sets.

- The median of a numerical data set is a measure of central tendency for a set of one-variable data.

- The *first quartile* of a data set is the median of the lower half of the data set.

- The *third quartile* is the median of the upper half of the data set.

- The *range*, the difference of the maximum and minimum values of a numerical data set, is a measure of dispersion.

All of these measures can provide a great deal of information about the data. A useful statistical display that combines the median, first quartile, third quartile, and the range is called a *box-and-whisker plot*. In a box-and-whisker plot, the data between the first and third quartiles is represented by a box. Inside the box, a line segment indicates the location of the median of the data. The data between the first quartile and the minimum and the data between the third quartile and the maximum are represented by line segments called whiskers.

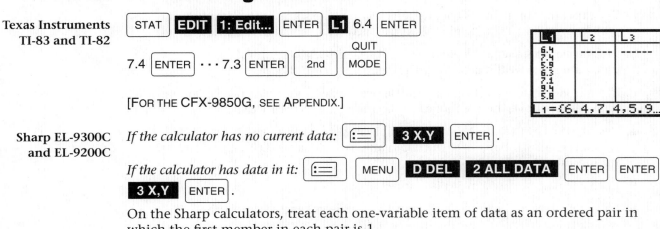

On the TI-82, Sharp EL-9300C and EL-9200C, and the HP 38G, you can represent one-variable data pictorially in a box-and-whisker plot. Begin by entering the data editor, then entering the data, and finally selecting the option that draws the box-and-whisker plot. Consider the following one-variable data set.

{6.4, 7.4, 5.9, 6.3, 7.1, 9.4, 5.8, 2.9, 6.4, 7.4, 6.4, 7.3}

Skill 1: Entering the Data

Texas Instruments TI-83 and TI-82

| STAT | **EDIT** | **1: Edit...** | ENTER | **L1** | 6.4 | ENTER |

QUIT

7.4 | ENTER | ··· 7.3 | ENTER | 2nd | MODE |

[FOR THE CFX-9850G, SEE APPENDIX.]

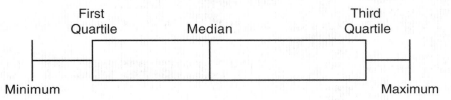

Sharp EL-9300C and EL-9200C

If the calculator has no current data: | ☷ | **3 X,Y** | ENTER |.

If the calculator has data in it: | ☷ | MENU | **D DEL** | **2 ALL DATA** | ENTER | ENTER |
3 X,Y | ENTER |.

On the Sharp calculators, treat each one-variable item of data as an ordered pair in which the first member in each pair is 1.

1 | ENTER | 6.4 | ENTER | 1 | ENTER | 7.4 | ENTER | ··· 1 | ENTER | 7.3 | ENTER |

Hewlett Packard HP 38G

| LIB | **Statistics** | **START** | **1 VAR ■** | ◄— If the display shows **2 VAR ■** press it.
If the display shows **1 VAR ■**,
6.4 | ENTER | 7.4 | ENTER | ··· 7.3 | ENTER | proceed to the next step.

Now that the data is entered into the calculator, you are ready to create the box-and-whisker plot.

- Be sure to clear any previously entered functions from the function list.

- If you are using a Texas Instruments TI-82 or TI-83, pressing $\boxed{\text{ZOOM}}$

  $\boxed{\text{9: Zoom Stat}}$ will calculate the ranges automatically. On a Sharp or Hewlett Packard calculator, the viewing window needed will automatically be determined.

Skill 2: Creating the Box-and-Whisker Plot

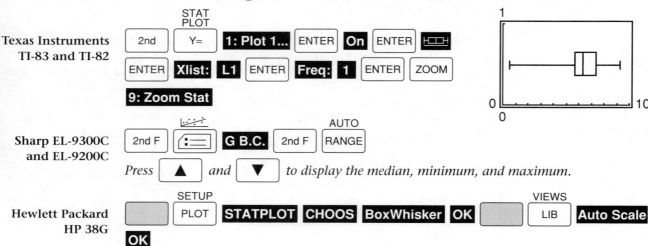

Texas Instruments TI-83 and TI-82

STAT PLOT

$\boxed{\text{2nd}}$ $\boxed{\text{Y=}}$ $\boxed{\text{1: Plot 1...}}$ $\boxed{\text{ENTER}}$ $\boxed{\text{On}}$ $\boxed{\text{ENTER}}$ $\boxed{\text{⊢□⊣}}$

$\boxed{\text{ENTER}}$ $\boxed{\text{Xlist:}}$ $\boxed{\text{L1}}$ $\boxed{\text{ENTER}}$ $\boxed{\text{Freq:}}$ $\boxed{\text{1}}$ $\boxed{\text{ENTER}}$ $\boxed{\text{ZOOM}}$

$\boxed{\text{9: Zoom Stat}}$

Sharp EL-9300C and EL-9200C

$\boxed{\text{2nd F}}$ $\boxed{\text{:≡}}$ $\boxed{\text{G B.C.}}$ $\boxed{\text{2nd F}}$ $\boxed{\text{RANGE}}$ AUTO

Press $\boxed{\blacktriangle}$ *and* $\boxed{\blacktriangledown}$ *to display the median, minimum, and maximum.*

Hewlett Packard HP 38G

SETUP

$\boxed{}$ $\boxed{\text{PLOT}}$ $\boxed{\text{STATPLOT}}$ $\boxed{\text{CHOOS}}$ $\boxed{\text{BoxWhisker}}$ $\boxed{\text{OK}}$ $\boxed{}$ $\boxed{\text{LIB}}$ VIEWS $\boxed{\text{Auto Scale}}$

$\boxed{\text{OK}}$

INVESTIGATIONS

1. Refer to the data set used in Skills 1 and 2. This list is the same set of data but arranged in ascending order for your convenience. The median of the data set is 6.4.

   {2.9, 5.8, 5.9, 6.3, 6.4, 6.4, 6.4, 7.1, 7.3, 7.4, 7.4, 9.4}

   a. Try experimenting with what happens if you make some of the data values a little lower and a little higher than the median all equal to one another.

   b. Then investigate what happens to the box-and-whisker plot if all the data values below the median are all made equal to the median.

   c. Then investigate what happens to the box-and-whisker plot if all the data values above the median are all made equal to the median.

   d. Restore the data set to the data set originally entered in Skill 1. Try finding out what will happen if the data value 9.4 is changed to 19.4 or the minimum data value 2.9 is changed to 0.

2. Refer to the data set used in Skills 1 and 2.

   a. Look for ways to make the length of the box small and each of the whiskers long. Try to describe how you did it and what characteristics the data set has after your efforts.

   b. Look for ways to make the length of the box large and each of the whiskers small. Try to describe how you did it and what characteristics the data set has after your efforts.

   c. Consider different ways to modify the data set so that it loses its whiskers in a box-and-whisker plot. In that case, consider what happens to the box.

GRAPHICS CALCULATOR SKILLS

GRAPHICS CALCULATOR SKILLS

Scatter Plots, Regression Equations, and Correlation Coefficients

It is frequently the case in real life that two quantities or variables have a definite relationship. For example, the length and weight of the blue whale share a relationship that can be modeled by a linear equation. The altitude of a rocket is related to the elapsed time of flight. Altitude actually depends on launch angle, initial velocity, air resistance, and gravity. Nonetheless, there is a quadratic functional relationship between altitude and elapsed flight time.

In many situations, you will be able to discern a pattern or functional relationship in a set of two-variable data by picturing the data in a *scatter plot*, a graph consisting of discrete points corresponding to the data pairs. If the points that represent the data pairs cluster in some recognizable way, then you may be able to look for an equation in two variables that reasonably represents the data. What follows is a description of how to

- enter two-variable data,

- display a scatter plot for that data,

- and model the data by means of a linear function.

Suppose you want to explore this short two-variable data set.
$$\{(0, 3.5), (1, 3.9), (2, 4.2), (3, 5.8), (4, 6.0), (5, 6.5)\}$$

Shown in Skill 1 is how to enter the data. Be sure to clear any previously entered data, functions, and graphs from the calculator. Use the down arrow to highlight the indicated menu choice, such as **1: Edit...**, and press ENTER.

Skill 1: Entering the Data

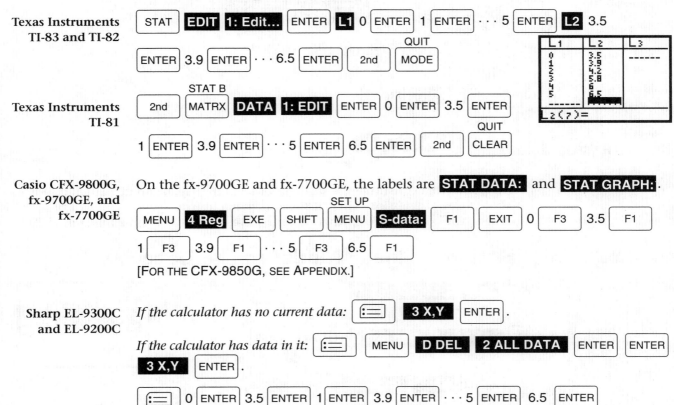

Texas Instruments TI-83 and TI-82
STAT **EDIT** **1: Edit...** ENTER **L1** 0 ENTER 1 ENTER ··· 5 ENTER **L2** 3.5
ENTER 3.9 ENTER ··· 6.5 ENTER 2nd MODE *(QUIT)*

Texas Instruments TI-81
2nd MATRX *(STAT B)* **DATA** **1: EDIT** ENTER 0 ENTER 3.5 ENTER
1 ENTER 3.9 ENTER ··· 5 ENTER 6.5 ENTER 2nd CLEAR *(QUIT)*

Casio CFX-9800G, fx-9700GE, and fx-7700GE
On the fx-9700GE and fx-7700GE, the labels are **STAT DATA:** and **STAT GRAPH:**.
MENU **4 Reg** EXE SHIFT MENU *(SET UP)* **S-data:** F1 EXIT 0 F3 3.5 F1
1 F3 3.9 F1 ··· 5 F3 6.5 F1
[FOR THE CFX-9850G, SEE APPENDIX.]

Sharp EL-9300C and EL-9200C
If the calculator has no current data: [≔] **3 X,Y** ENTER.
If the calculator has data in it: [≔] MENU **D DEL** **2 ALL DATA** ENTER ENTER
3 X,Y ENTER.
[≔] 0 ENTER 3.5 ENTER 1 ENTER 3.9 ENTER ··· 5 ENTER 6.5 ENTER

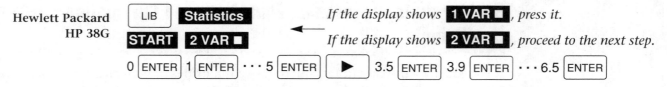

Hewlett Packard
HP 38G

LIB **Statistics**
START 2 VAR ■

If the display shows **1 VAR ■** *, press it.*

If the display shows **2 VAR ■** *, proceed to the next step.*

0 ENTER 1 ENTER ⋯ 5 ENTER ▶ 3.5 ENTER 3.9 ENTER ⋯ 6.5 ENTER

Now that you have entered the data into the calculator, you can create the scatter plot. For this set of data, the viewing window values have been set at $0 \le x \le 6$ and $0 \le y \le 7$.

Skill 2: Creating the Scatter Plot

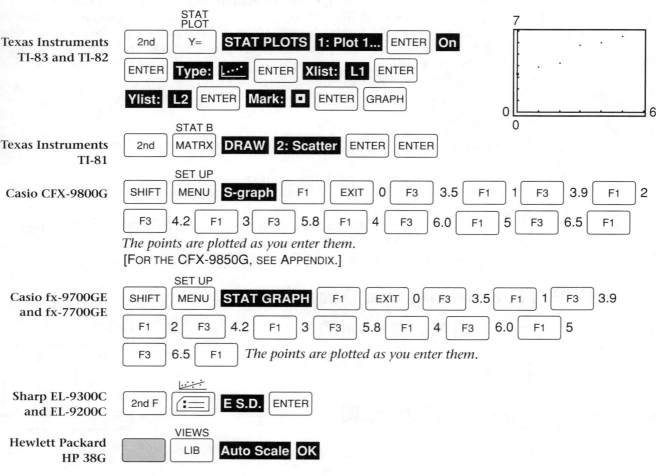

Texas Instruments
TI-83 and TI-82

2nd Y= (STAT PLOT) **STAT PLOTS 1: Plot 1...** ENTER **On**
ENTER **Type: ┗∙∙∙** ENTER **Xlist: L1** ENTER
Ylist: L2 ENTER **Mark: ▫** ENTER GRAPH

Texas Instruments
TI-81

2nd MATRX (STAT B) **DRAW 2: Scatter** ENTER ENTER

Casio CFX-9800G

SHIFT MENU (SET UP) **S-graph** F1 EXIT 0 F3 3.5 F1 1 F3 3.9 F1 2
F3 4.2 F1 3 F3 5.8 F1 4 F3 6.0 F1 5 F3 6.5 F1

The points are plotted as you enter them.
[FOR THE CFX-9850G, SEE APPENDIX.]

Casio fx-9700GE
and fx-7700GE

SHIFT MENU (SET UP) **STAT GRAPH** F1 EXIT 0 F3 3.5 F1 1 F3 3.9
F1 2 F3 4.2 F1 3 F3 5.8 F1 4 F3 6.0 F1 5
F3 6.5 F1 *The points are plotted as you enter them.*

Sharp EL-9300C
and EL-9200C

2nd F ▭ **E S.D.** ENTER

Hewlett Packard
HP 38G

☐ LIB (VIEWS) **Auto Scale OK**

The points that constitute the scatter plot seem to indicate an upward and somewhat steady progression. Thus, it suggests that a line might pass through or be close to the points representing the data. Here is how to obtain a linear model.

Skill 3 : Fitting a Linear Model to the Data

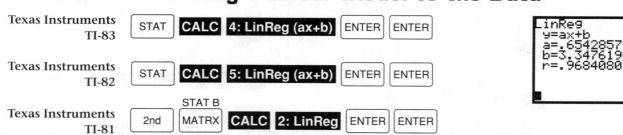

Texas Instruments
TI-83

STAT **CALC 4: LinReg (ax+b)** ENTER ENTER

Texas Instruments
TI-82

STAT **CALC 5: LinReg (ax+b)** ENTER ENTER

Texas Instruments
TI-81

2nd MATRX (STAT B) **CALC 2: LinReg** ENTER ENTER

```
LinReg
 y=ax+b
 a=.6542857143
 b=3.347619048
 r=.9684080101
```

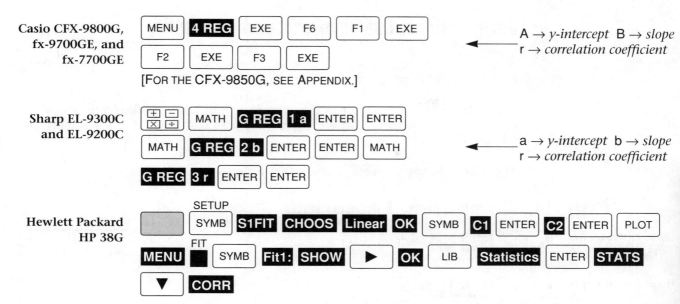

Casio CFX-9800G, fx-9700GE, and fx-7700GE

MENU | 4 REG | EXE | F6 | F1 | EXE

F2 | EXE | F3 | EXE

[FOR THE CFX-9850G, SEE APPENDIX.]

← A → *y-intercept* B → *slope*
r → *correlation coefficient*

Sharp EL-9300C and EL-9200C

[+ −] [× ÷] | MATH | G REG | 1 a | ENTER | ENTER

MATH | G REG | 2 b | ENTER | ENTER | MATH

G REG | 3 r | ENTER | ENTER

← a → *y-intercept* b → *slope*
r → *correlation coefficient*

Hewlett Packard HP 38G

[ ] SETUP | SYMB | S1FIT | CHOOS | Linear | OK | SYMB | C1 | ENTER | C2 | ENTER | PLOT

MENU | FIT | SYMB | Fit1: | SHOW | ▶ | OK | LIB | Statistics | ENTER | STATS

▼ | CORR

The correlation coefficient of a set of two-variable data is a measure of how well the data variables are related to one another. Each of the calculators will report a value for the correlation coefficient *r*. Since *r* is close to 1, the linear fit for the data is quite good.

INVESTIGATIONS

1. Try creating scatter plots for some of the data sets that follow. Some patterns can be discerned quickly from the data before you make the scatter plot.

Table A (x, y)	Table B (x, y)	Table C (x, y)	Table D (x, y)
(0, 12.5)	(0, 13.1)	(0, 13.4)	(0, 11.3)
(1, 12.5)	(1, 12.9)	(1, 13.1)	(1, 11.7)
(2, 12.5)	(2, 12.7)	(2, 12.8)	(2, 12.1)
(3, 12.5)	(3, 12.5)	(3, 12.5)	(3, 12.5)
(4, 12.5)	(4, 12.3)	(4, 12.8)	(4, 12.1)
(5, 12.5)	(5, 12.1)	(5, 13.1)	(5, 11.7)
(6, 12.5)	(6, 11.9)	(6, 13.4)	(6, 11.3)

2. Consider again the data entered in Skill 1 and displayed in a scatter plot in Skill 2.

   a. Investigate what you would get if you chose a quadratic model instead.

   b. Investigate what you would get if you chose an exponential model instead.

3. Try another experiment with the data used in Skills 1 and 2.

   a. Try exploring what happens if each of the *x*-values in the data set is kept as is but each of the *y*-values in the data set is doubled. [The slope and *y*-intercept of the line of best fit double. However, the correlation coefficient remains the same as before.]

   b. Try exploring what happens if each of the *x*-values in the data set is kept as is but each of the *y*-values in the data set is halved. [The slope and *y*-intercept of the line of best fit are halved. However, the correlation coefficient remains the same as before.]

   c. Try exploring what happens if each of the *x*-values in the data set is kept as is but each of the *y*-values in the data set is increased by 1. [The slope and correlation coefficient remain the same as before. However the *y*-intercept is increased by 1.]

GRAPHICS CALCULATOR SKILLS

Permutations, Combinations, and Probability

Many counting and probability problems involve factorials, permutations, combinations, and random integer generators.

For example, in how many ways can the first eight letters of the English alphabet be arranged to form sensible and nonsensible words? The answer is 8 factorial, denoted 8!.

The following key sequences illustrate the computation of 8!.

Skill 1: Computing Factorials

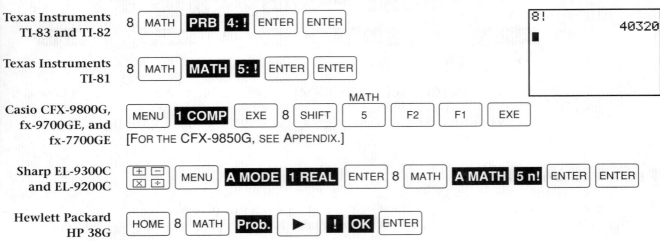

In how many ways can the first eight letters of the English alphabet be arranged into words if a word consists of exactly five letters? The answer is the number of permutations of eight distinct objects taken five at one time. The actual calculation is shown in the following key sequences.

Skill 2: Computing Permutations

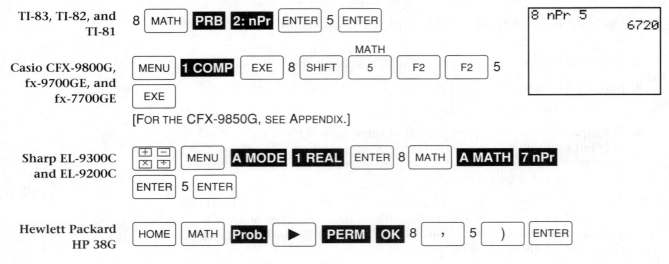

In how many ways can the first eight letters of the English alphabet be arranged into five-letter groups? The answer is the number of combinations of eight distinct objects taken five at one time. The actual calculation is shown in this set of key sequences.

Skill 3: Computing Combinations

TI-83, TI-82, and TI-81

8 | MATH | **PRB** | **3: nCr** | ENTER | 5 | ENTER

Casio CFX-9800G, fx-9700GE, and fx-7700GE

MENU | **1 COMP** | EXE | 8 | SHIFT | 5 (MATH) | F2 | F3 | 5 | EXE

[FOR THE CFX-9850G, SEE APPENDIX.]

Sharp EL-9300C and EL-9200C

[+ −] [× ÷] | MENU | **A MODE** **1 REAL** | ENTER | 8

MATH | **A MATH** **6 nCr** | ENTER | 5 | ENTER

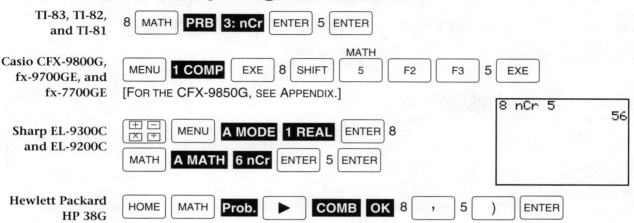

```
8 nCr 5
                    56
```

Hewlett Packard HP 38G

HOME | MATH | **Prob.** | ▶ | **COMB** **OK** | 8 | , | 5 |) | ENTER

In many probability simulations, you will want to randomly generate positive integers from a set such as {1, 2, 3, 4, 5, 6}. To accomplish this, you will need to evaluate the function stated here.

roll = [6 × random number between 0 and 1] + 1

Skill 4: Generating a Random Integer From a Set

Texas Instruments TI-83

MATH | **PRB** **5: rand Int(** | 1 | , | 6 |) | ENTER

```
Int (6Rand)+1
                    6
```

Texas Instruments TI-82

MATH | **NUM** **4: Int** | ENTER | (| 6 | MATH | **PRB** **1: rand**

ENTER |) | + | 1 | ENTER

Texas Instruments TI-81

MATH | **NUM** **4: Int** | ENTER | (| 6 | MATH | **PRB** **1: Rand** | ENTER |)

+ | 1 | ENTER

Casio CFX-9800G, fx-9700GE, and fx-7700GE

MENU | **1 COMP** | EXE | SHIFT | 5 (MATH) | F3 | F2 | EXIT | (| 6 | F2

F4 |) | + | 1 | EXE

[FOR THE CFX-9850G, SEE APPENDIX.]

Sharp EL-9300C and EL-9200C

[+ −] [× ÷] | MENU | **A MODE** **1 REAL** | ENTER | MATH | **A MATH** **2 int** | ENTER

(| 6 | MATH | **A MATH** **8 random** | ENTER |) | + | 1 | ENTER

Hewlett Packard HP 38G

HOME | MATH | **Real** | ▶ | **FLOOR** **OK** | 6 | MATH | **Prob.** | ▶ | **RANDOM** **OK**

) | + | 1 | ENTER

Shown on this page are the keystrokes needed to find the probability that five cards drawn at random from a standard deck of fifty-two cards will contain exactly two clubs. The keystrokes that follow exemplify the calculation of $\dfrac{_{13}C_2 \cdot {_{39}C_3}}{_{52}C_5}$.

Skill 5: Finding Probabilities Involving Combinations

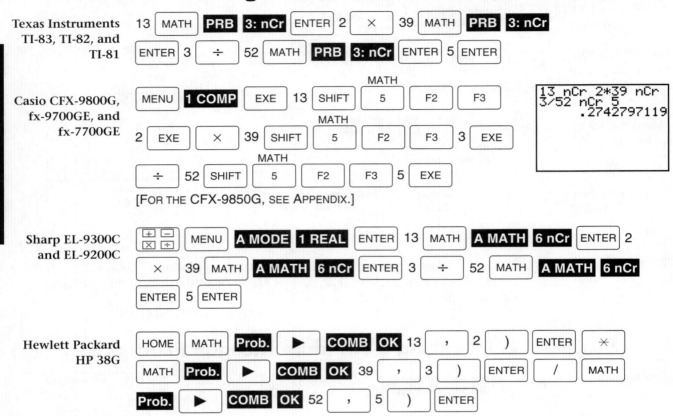

[FOR THE CFX-9850G, SEE APPENDIX.]

INVESTIGATIONS

1. **a.** Try computing 4!, 4.5!, and (–4.5)!. What does the display indicate?
   [When trying to compute the second and third expressions, an error message appears. The factorial operation is only defined for nonnegative integers.]

   b. Try computing $_6P_3$, $_3P_6$, $_6C_3$, and $_3C_6$. What does the display indicate?
   [In both $_nP_r$ and $_nC_r$, the value of n must be greater than or equal to the value of r otherwise an error message appears.]

   c. The expression $\dfrac{n!}{(n-r)!}$ gives the number of permutations of n distinct

   objects taken r at a time. Use the calculator to evaluate this expression for different choices of n and r. Confirm that it gives the same number as $_nP_r$.

2. Suppose that the concept of a number cube is generalized so that any *regular polyhedron* becomes a number polyhedron. (There are five regular polyhedra. They are *tetrahedron* (four congruent equilateral triangular faces), *cube* (six congruent square faces), *octahedron* (eight congruent equilateral triangular faces), *icosahedron* (twenty congruent equilateral triangular faces), and *dodecahedron* (twelve congruent regular pentagonal faces).

   a. Explore random integer functions that simulate the roll of a regular polyhedron.

   b. Set up and investigate some probability experiments involving these solids.

Introduction to Spreadsheet Software in the Classroom

Perhaps the most common computer application used in the classroom is the spreadsheet. When many people hear the word *spreadsheet*, they think of accountants and statisticians, people who spend a great deal of time number crunching. While it is true that spreadsheets are used in accounting, statistical studies, and related endeavors, one must not make the mistake of thinking that the spreadsheet is restricted to only those fields.

A great deal of mathematical activity involves the use of formulas, functions, and mathematical objects defined by them. Since the spreadsheet has the capability of evaluating mathematical expressions for input values given in a variety of ways, the spreadsheet can be invaluable in exploring algebraic and numerical patterns. It should also be pointed out that a spreadsheet is a vast array of cells, perhaps 256 columns and as many rows. Thus, the user can tackle multiple or extensive problems.

Perhaps the most powerful feature of a spreadsheet is its ability to update and recalculate when input data are changed. Many people call this capability the "what if" capability of the spreadsheet. Here are two examples of what this means.

- Suppose that you wish to deposit P dollars into a bank account for which the annual interest is $r\%$, the interest is compounded yearly, and the money is left in the account for t years. Using the spreadsheet, you can enter a fixed amount for P, enter a fixed value of r, and calculate the amount A in the bank for variable values of t. What if the amount deposited is changed to a different number? Calculation of A for various values of t is accomplished with one simple change in one of the spreadsheet's cells. All the manual labor is removed. The comparison of financial growth is now quickly accessible. The same comments apply if P is kept the same but r is changed to a different number.

- Suppose that you wish to study how the area A of a square changes as the length s of a side changes. You can use the spreadsheet to enter various lengths s of sides and have it calculate the area A of the square. It is easy to see from the spreadsheet table that if s is doubled, A is not doubled but rather is quadrupled. What if you want to investigate how the area of a regular pentagon varies with the length s of a side? Simply replace the area formula $A = s^2$ with the pentagon's area formula, $A = 1.72048s^2$.

Another important feature of a spreadsheet is that of charting. You can easily chart data, whether statistical or functional, in a spreadsheet chart. A chart representing quadratic data is shown here.

The software program featured in this book is *Microsoft® Excel* (version 3.0). Different spreadsheets are filled out in essentially the same way. Consequently, what is described in the pages that follow can be adapted fairly easily to spreadsheets developed and produced by other vendors.

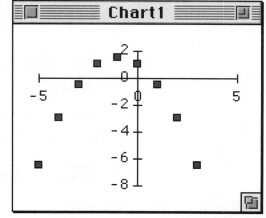

Spreadsheet Basics

You can open, or launch, a spreadsheet application by highlighting the spreadsheet application name or icon or by highlighting the name or icon of a particular spreadsheet document. Then click on it twice.

To open a particular spreadsheet document when the spreadsheet application has already been launched, follow these steps.

- Click on **File**, then **Open...**, in the main menu at the top of the computer display.

- In the list of documents that appears, use the mouse pointer to highlight the name of the document you want. Click on **Open...**.

This diagram illustrates the components of the worksheet that appears.

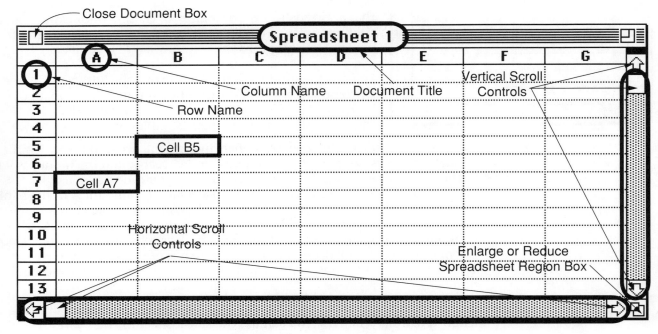

Above the worksheet, you will see the entry bar. It identifies the cell where your cursor is located. As you type into it, this display will show what you have typed.

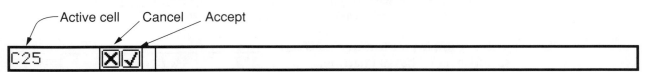

To save a spreadsheet document you have created, follow these steps.

- Select **File**, then **Save As...**. Type a name for your document. Click on **Save**.

When you have finished working on a particular worksheet, follow these steps.

- To close the spreadsheet document, click on the small square at the upper left corner of the document. To close the spreadsheet application, click on **File**, then **Quit**.

Spreadsheet Text, Formulas, and Charts

When you create a spreadsheet, you use text, formulas, and charts.

To enter text into a particular cell, follow these steps.

- Highlight the cell where you want the text to begin.
- Use the keyboard as you would use a typewriter to enter your text.
- In the entry bar, click on ✓.

To enter a formula into a particular cell, follow these steps.

- Highlight that cell.
- Type =. Then type the formula using the notation commonly found in software programs. Click on ✓.

$\begin{cases} + & \text{for addition} \\ - & \text{for subtraction} \\ * & \text{for multiplication} \\ / & \text{for division} \\ \char`^ & \text{for exponentiation} \end{cases}$

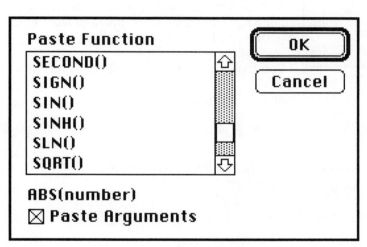

Paste Function

| SECOND() |
| SIGN() |
| SIN() |
| SINH() |
| SLN() |
| SQRT() |

OK

Cancel

ABS(number)

⊠ **Paste Arguments**

To enter a built-in function like *sine* and *square root*, follow these steps.

- Select **Formula** from the main menu.
- Select **Paste Function. . .**.
- From the dialog box that appears, scroll down to the desired function name and highlight it.
- Click on **OK**.
- In the entry bar at the top of the spreadsheet, fill in the arguments needed by the function.
- Click on ✓.

To edit the contents of a particular cell, follow these steps.

- Place the mouse pointer on that cell.
- In the entry bar, type your changes.
- Click on ✓.

To create a chart of the data in your spreadsheet, follow these steps.

- Highlight the data you wish to chart.
- Select **File**, then **New. . .**, then **Chart**. Click on **OK**. If your data contains values of *x* and corresponding values of *y*, click on **X-Values for XY-Chart**. Click on **OK**.

Once the chart appears in its window, you may choose a different chart type.

- Select **Gallery** from the main menu.
- Select **Line. . .**, **Bar. . .**, or some other chart type.
- Highlight the style you desire. Click on **OK**.

Averages

To see how you can use a spreadsheet both for your own needs and for those of your students, complete the spreadsheet activity outlined below.

	A	B	C	D	E	F	G
	Grades						
1	Student Test Grades						
2		A	B	C	D	E	
3	Test No.						
4	1	80	73	90	77	88	
5	2	82	74	89	78	86	
6	3	84	78	92	79	85	
7	4	78	79	92	73	88	
8	5	86	73	93	72	82	
9							
10							
11							

- In cell A1, type **Student Test Grades**.

- Click on cell B2. Type A. Press [→] or [Tab] to advance to cell C2. Type B. Enter C, D, and E into cells D2, E2, and F2.

- In cell A3, type **Test No**. In cells A4 through A8, type 1, 2, 3, 4, and 5. (To advance from one cell to the cell directly below, press [Return] or [↓].)

- Type the test grades into cells B4 through B8, C4 through C8, D4 through D8, E4 through E8, and F4 through F8.

You have now set up a small grade sheet.

Suppose that you want to find Student A's average grade for the five tests.

- In cell B9, type =SUM(B4:B8)/5. Click on ✓ in the entry bar.

Suppose you want to find the average of the grades for Test No. 1.

- In cell G4, type =SUM(B4:F4)/5. Click on ✓ in the entry bar. The average grade of the five students on Test No. 1 will appear in cell G4.

Finally, you may want to find the class average over the entire grading period.

- Select a cell such as A10. Type =SUM(B4:F8)/25. Click on ✓ in the entry bar.

You may have noticed that the columns in the spreadsheet are somewhat narrow. To change the width of a group of consecutive columns, follow these steps.

- Highlight the columns whose width you want to change.

- From the main menu, select **Format**, then **Column Width...**.

- The larger the number you enter into the dialog box, the wider the columns will become. The smaller the number, the narrower the columns will become. Click on **OK**.

To center cell contents in a row or column:

- Highlight the cells whose entries you wish to center.

- Select **Format**, **Alignment...**, **Center**. Click on **OK**.

Formulas Involving Two Variables

One of the simplest formulas from geometry is the formula for the area of a rectangle. If a rectangle has length *x* units and width *y* units, then the area is the product of *x* and *y*. The spreadsheet shown gives the area of a rectangle whose length varies from 1 to 25 in increments of 1 and whose width varies from 1 to 10 in increments of 1. The following is an efficient way to create the table. In cell A1, type X/Y.

Create row 1 as follows.

- In cell B1, enter 1.
- In cell C1, enter =1+B1.
- Highlight Cells C1 through K1.
- Select **Edit**, then **Fill Right**.

Create column A as follows.

- In cell A2, enter 1.
- In cell A3, enter =1+A2.
- Highlight cells A3 through A26.
- Select **Edit**, then **Fill Down**.

Using $, you hold one variable constant and let the second variable take on different values.

Create the body of the table as follows.

- In cell B2, enter =$A2*B$1.
- Highlight cells B2 through B26.
- Select **Edit**, then **Fill Down**.
- Highlight cells B2 through K26.
- Select **Edit**, then **Fill Right**.

	A	B	C	D	E	F	G	H	I	J	K
1	X/Y	1	2	3	4	5	6	7	8	9	10
2	1	1	2	3	4	5	6	7	8	9	10
3	2	2	4	6	8	10	12	14	16	18	20
4	3	3	6	9	12	15	18	21	24	27	30
5	4	4	8	12	16	20	24	28	32	36	40
6	5	5	10	15	20	25	30	35	40	45	50
7	6	6	12	18	24	30	36	42	48	54	60
8	7	7	14	21	28	35	42	49	56	63	70
9	8	8	16	24	32	40	48	56	64	72	80
10	9	9	18	27	36	45	54	63	72	81	90
11	10	10	20	30	40	50	60	70	80	90	100
12	11	11	22	33	44	55	66	77	88	99	110
13	12	12	24	36	48	60	72	84	96	108	120
14	13	13	26	39	52	65	78	91	104	117	130
15	14	14	28	42	56	70	84	98	112	126	140
16	15	15	30	45	60	75	90	105	120	135	150
17	16	16	32	48	64	80	96	112	128	144	160
18	17	17	34	51	68	85	102	119	136	153	170
19	18	18	36	54	72	90	108	126	144	162	180
20	19	19	38	57	76	95	114	133	152	171	190
21	20	20	40	60	80	100	120	140	160	180	200
22	21	21	42	63	84	105	126	147	168	189	210
23	22	22	44	66	88	110	132	154	176	198	220
24	23	23	46	69	92	115	138	161	184	207	230
25	24	24	48	72	96	120	144	168	192	216	240
26	25	25	50	75	100	125	150	175	200	225	250

You can also use this spreadsheet to explore factors of whole numbers. If, for example, you locate 12 in the body of the table, you can see that it is the product of 1 and 12, 12 and 1, 2 and 6, 6 and 2, and so on.

You can modify the spreadsheet to study the distance traveled in *x* minutes at the rate of *y* miles per hour by changing the formula in cell B2 from =$A2*B$1 to =$A2*B$1/60. Refill the body of the table as described.

To show how an investment of $1000 grows over different numbers of years and at different annual interest rates, use column A for the number of years and row 1 for different annual interest rates, such as *r* = 0.04, 0.045, 0.050, ... In cell B1, enter 0.04. In cell C1, enter =0.005+B1. **Fill Right**. In cell B2, enter =ROUND(1000*(1+B$1)^$A2,2). Refill the body of the table as described.

Creating a Student Handout

There are many times when you may wish to make a student handout. You can use a spreadsheet to accomplish your goal. On the next page, you will see a sample handout created in *Microsoft® Excel*. To create it, follow these steps.

To create text, such as the title:

- Place the cursor in cell A1. Type A Study in Projectile Motion.

- Highlight cell A1. Select **Format** from the main menu. Select **Font...**. In the dialog box that appears, click on a font like Helvetica and a size such as 24. Click on the box for bold.

- Click on **OK**. Notice that the title fills as many cells in row 1 as it needs.

In cells A2 through A4, enter the text shown line-for-line.

Fill in cells A7 through B21 with the numbers shown on the next page. To create a chart or graph of the data, follow the instructions previously given.

To bring your chart into the spreadsheet, proceed as follows.

- Make the chart window active by clicking on a visible portion of it.

- From the main menu, select **Chart**.

- From the submenu that appears, select **Select Chart**.

- Copy the chart by selecting **Edit**, then **Copy**.

- Make the spreadsheet active by clicking on a visible portion of it.

- Paste the chart into the spreadsheet by selecting **Edit**, then **Paste**.

To position and size the chart:

- Reposition the chart by placing the mouse pointer on it and dragging the chart to a different location.

- With the chart selected, place the mouse pointer on the lower right corner ■ to resize in both horizontal and vertical directions.

To enter text, such as that found in the questions, use the keyboard as a typewriter. Estimate the amount of copy you can fit on one line. Press **Return**.

To change row height or column width, follow these steps.

- Highlight the row(s) or column(s) whose height or width you wish to change.

- Select **Format**, then **Row Height...** or **Column Width...**. Change the height or width number. Click on **OK**.

To print your completed handout, carry out each of these steps.

- Highlight the rows and columns that constitute your handout. From the main menu, select **Options**, then **Set Print Area**.

- From the main menu, select **File**, then **Print...**. Click on **Print**.

The illustration on this page was created using the instructions from the preceding page.

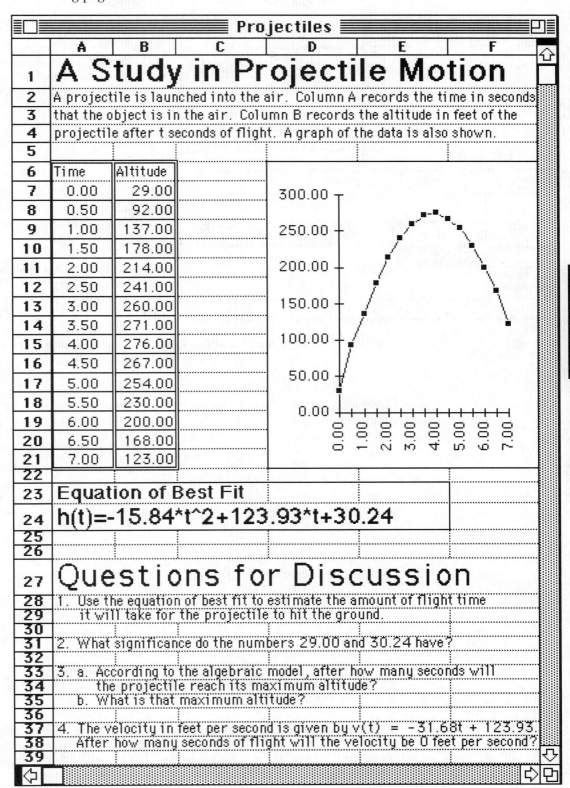

Projectiles

A Study in Projectile Motion

A projectile is launched into the air. Column A records the time in seconds that the object is in the air. Column B records the altitude in feet of the projectile after t seconds of flight. A graph of the data is also shown.

Time	Altitude
0.00	29.00
0.50	92.00
1.00	137.00
1.50	178.00
2.00	214.00
2.50	241.00
3.00	260.00
3.50	271.00
4.00	276.00
4.50	267.00
5.00	254.00
5.50	230.00
6.00	200.00
6.50	168.00
7.00	123.00

Equation of Best Fit

$$h(t)=-15.84*t^2+123.93*t+30.24$$

Questions for Discussion

1. Use the equation of best fit to estimate the amount of flight time it will take for the projectile to hit the ground.

2. What significance do the numbers 29.00 and 30.24 have?

3. a. According to the algebraic model, after how many seconds will the projectile reach its maximum altitude?
   b. What is that maximum altitude?

4. The velocity in feet per second is given by $v(t) = -31.68t + 123.93$. After how many seconds of flight will the velocity be 0 feet per second?

Introduction to *f(g) Scholar*™ in the Classroom

One of the issues that teachers using technology must struggle with is this: "Which technology might I choose to handle today's topic? Do I use a spreadsheet or a calculator? Do I use tables of numbers and graphs to accompany them?"

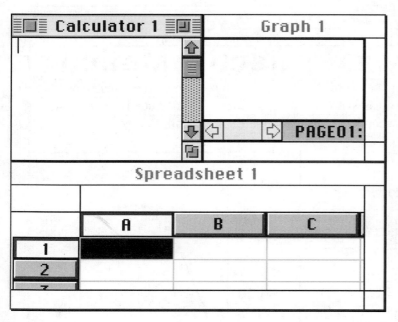

The software application known as *f(g) Scholar*™ is a recent addition to the array of software programs available, and it can help teachers resolve this issue. The application combines three resources:

- a calculator,

- a graph window,

- and a spreadsheet.

Consequently, one can work completely within *f(g) Scholar*™ and handle formulas, functions, tables, sequences, matrices, graphs, data, statistical measures, and more.

On the pages that follow, you will be introduced to how *f(g) Scholar*™ works, and you will also see it in action as applied to some mathematical problems. The final pair of pages illustrates how *f(g) Scholar*™ can be used to create a student handout. Once you work through the steps involved, you might decide to challenge students with activities not covered in this book.

Use of the *f(g) Scholar*™ calculator requires some special instruction since it behaves a bit differently than a graphics calculator. If you are familiar with spreadsheets, you will find that the *f(g) Scholar*™ spreadsheet behaves much like other spreadsheet programs.

When you launch, or open, *f(g) Scholar*™, you will notice that it has a main menu and toolbar at the top of the computer display. The items available depend on the window you choose to make active. For example, if the spreadsheet window is active and you click on the calculator window, you will see a different toolbar become available at the top of the computer display. Space allows the explanation of only some of the items in the main menu and toolbar. Once familiar with the navigation of these, you find the use of other menu and toolbar items manageable.

Because *f(g) Scholar*™ combines a calculator, a graph window, and a spreadsheet, you will see three working windows appear when you start the program. You can have one or more windows available to you at any given time. Furthermore, you may close windows that you do not need and enlarge the window that you want to fill the computer display. This step, along with the program's ability to show larger type, will help students in all parts of the classroom clearly see what is in your demonstration.

As in the use of any software program, you will discover that there are advanced features tucked away in menus and that there are subsidiary menus that you can access from them. Do not let these advanced features intimidate you. Your use of *f(g) Scholar*™ will be most effective if you plan well-defined and straightforward demonstrations at the start and then try more sophisticated demonstrations as the school year progresses. Good luck in using *f(g) Scholar*™.

Working With *f(g) Scholar*™ Windows

When you launch, or open, $f(g)$ *Scholar*™, you will see three application windows on your computer display as shown. The window with the dark border is the active window. In the diagram on this page, the spreadsheet window is active.

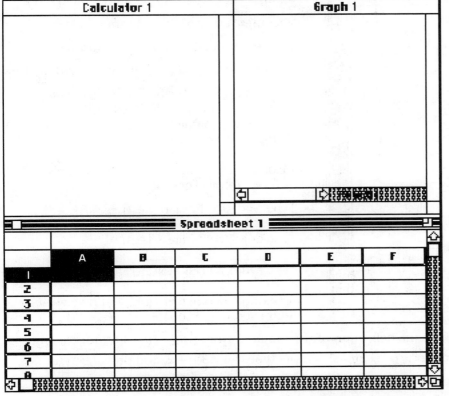

- To use a different window, click on that window. The border of the selected window will then darken. The borders of the other windows will become dim.

You may wish to work only in one or two windows.

- To close a particular window, click on the small square at the upper left of the window.

- To open a window that is closed, select **Windows** and then **New** from the main menu at the top of the screen. Then choose the window you wish to open by clicking on its name.

At times, you may wish to work in a larger or smaller window.

- To change the size of a window, place the mouse pointer on the square at the lower right of the window. With the mouse button depressed, drag the corner to enlarge or reduce the window. (You will see the window change size as you drag the corner.)

In $f(g)$ *Scholar*™, you can create or set up a classroom demonstration the day before you wish to use it. In the appropriate windows, enter the material you wish to use at the outset of the presentation. Then follow these steps.

- From the main menu, choose **File**, then **Save As...**.
- Give the document a name.
- Click on **Save**.
- Click on **Quit**.

When you wish to begin a demonstration, locate the document and simply double-click on its name or *icon* with the mouse pointer.

On the pages that follow, you will learn how to use

- the calculator,
- the spreadsheet,
- the grapher, and
- the calculator, spreadsheet, and grapher together.

Before continuing, examine various main-menu submenus to see what options and commands are available to you at different times.

Function Tables

Many functions are defined by equations. To enter a function defined by an equation into the *f(g) Scholar*™ calculator, follow these steps.

```
┌─────────────────────────────────────────┐
│ ▤□▭▬▬▬▬▬   Calculator 1      ▭  ▣         │
│                                    ⇧      │
│ n = 1, 10, 1                       ⇩      │
│ >N defined from 1 to 10 by 1              │
│ f(n)=2*n-3                                │
│ >Function F defined.                      │
│ ev(f(n))                                  │
│ 1            ->              -1           │
│ 2            ->               1           │
│ 3            ->               3           │
│ 4            ->               5           │
│ 5            ->               7           │
│ 6            ->               9           │
│ 7            ->              11           │
│ 8            ->              13           │
│ 9            ->              15   ⇧        │
│ 10           ->              17   ⇩        │
│                                   ▣       │
└─────────────────────────────────────────┘
```

- Enter the variable name, its starting value, its ending value, and a step, or increment value. Press ENTER from the computer keyboard.

- Type the function name in the form

  f (*variable name*) = *expression*. Press ENTER from the computer keyboard. The variable name must be the one you declared in the first step. The function name can be a letter such as f or a string of letters such as func.

For example, to evaluate $f(n) = 2n - 3$ if $n = 1, 2, 3, \ldots, 9, 10$, follow these steps.

- Type n=1,10,1. Press ENTER.

- Type f(n)=2*n – 3. Press ENTER.

- Type ev(f(n)). Press ENTER.

After exploring one function, you may wish to examine a different one.

- To change the definition of the function from $f(n) = 2n - 3$ to, for example, the function $f(n) = 3n + 2$, simply type f(n) = 3*n + 2, and press ENTER.

- Type ev(f(n)) and press ENTER. The computer will then make the necessary changes and display the new function table.

To display the function table for a current function and a new one, follow these steps.

- Place the cursor below the table.

- Type a new variable name, such as t, and the new function's domain. Press ENTER.

- Type a new function, with a name such as g, and press ENTER.

- Type ev(g(t)). Press ENTER.

To display two tables in separate calculator windows simultaneously, follow this procedure.

- From the main menu, select **Windows**, then **New ▶**, then **Calculator**.

- Follow the steps to enter the new function along with its variable name and domain.

- Resize and position the two calculator windows as desired.

Graphing Functions

Make sure the calculator and graph windows are open.

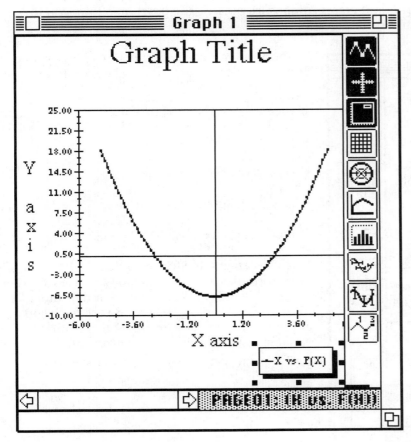

To graph a function such as $f(x) = x^2 - 7$, follow these steps.

- Activate the calculator window.
- Type x= −4,4,0.1. Press ENTER .
- Type f(x)=x^2−7. Press ENTER .
- Type graph(x,f(x)). Press ENTER .

Notice that the *x*- and *y*-axes appear at the left and bottom of the graph display. To show axes whose intersection is the origin:

- Click on the graph window.
- Click on ⊕ , the second tool in the vertical tool bar at the right of the graph window.

You can move certain pieces of the display with the mouse pointer by touching an edge of the item and dragging it to another location. In the display on this page, the legend ⊢X vs. F(X) has been moved to the bottom of the window.

To adjust the interval over which the function is graphed, carry out these steps.

- From the calculator or graph window, select **Graph** in the main menu.
- Select **Graph Scaling...** .
- Click on **Manual** . Set your minimum and maximum values for *x* and *y*.
- Click on **OK** .

Follow these steps to graph the following set of functions for $-4 \le x \le 4$.

$$a(x) = x^2 + 3 \qquad b(x) = x^2 \qquad c(x) = x^2 - 4$$

- Type x=−4,4,0.1 in the calculator window. Press ENTER from the computer keyboard.
- Type each function as already described. Press ENTER after typing each function.
- Type graph(x,a(x),b(x),c(x)). Press ENTER .

f(g) Scholar ™   Combinations of Functions

Many teachers have students graph composites of functions to see how their characteristics compare with those of the original functions. In *f(g) Scholar*™, you can easily form and graph composites because the program allows you to use function composition notation.

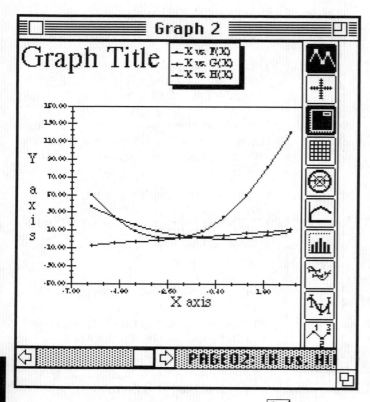

To form the composite of functions $f(x) = x^2$ and $g(x) = 2x + 5$ to make $f(g(x)) = (2x + 5)^2$, follow these steps.

- Type x= –5,5,1. Press ENTER .

- Type f(x)=x^2. Press ENTER .

- Type g(x)=2*x+5. Press ENTER .

- Type h(x)=f(g(x)). Press ENTER .

- Type graph(x,f(x),g(x),h(x)). Press ENTER .

You also can graph a function defined by two or more formulas. Use two variables, such as x for the first formula and t for the second formula. Make the domain of the second variable begin at the maximum value specified for the first variable. For example, consider this function:

$$y = \begin{cases} 2x + 3 & \text{if } 0 \le x \le 5 \\ |0.5x^2 - 10| & \text{if } 5 < x \le 10 \end{cases}$$

The keystrokes needed to graph y are shown.

You can select $\sqrt[2]{\mathsf{x}}$ to access algebraic functions such as the absolute value function.

You can select 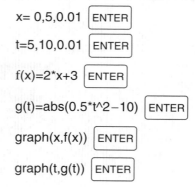 to access trigonometric functions. To insert a function from one of the menus:

x= 0,5,0.01 ENTER

t=5,10,0.01 ENTER

f(x)=2*x+3 ENTER

g(t)=abs(0.5*t^2–10) ENTER

graph(x,f(x)) ENTER

graph(t,g(t)) ENTER

- Place the cursor in the calculator where you wish the function to be placed.

- Click on one of the icons from the toolbar at the top of the computer display.

- Place the mouse pointer on ⬇ in the right border of the dialog box that appears. Depress the mouse button until you see the function you want.

- Highlight your selection. Click on **OK** .

To use a trigonometric function, select either degree or radian mode using **Calculator** , then **Angular Mode ▶** .

Trace and Zoom

Using trace and zoom features in *f(g) Scholar*™, you can explore the problem of finding solutions of a system of linear equations. Here is how you might proceed to solve the following system.

$$\begin{cases} 3x + 2y = 12 \\ x + y = 7 \end{cases} \longrightarrow \begin{cases} f(x) = \dfrac{12 - 3x}{2} \\ g(x) = 7 - x \end{cases}$$

- Activate the calculator window.

- Type x=−3.5,3.5,0.1. Press ENTER .

- Type f(x)=(12−3*x)/2. Press ENTER .

- Type g(x)=7−x. Press ENTER .

- Type graph(x,f(x),g(x)). Press ENTER .

- Click on the graph window.

- From the main toolbar at the top of the display, click on ⊹⊕⊹ . Crosshairs will appear.

- Place the mouse pointer on the point where the graphs intersect. Click once. The intersection of the crosshairs moves to that point, and the *x*- and *y*-coordinates of the point appear at the bottom of the computer display. You will get a good approximation of the solution.

To zoom in on the graph, follow these steps.

- Click on ⊹⊕⊹ , then 🔍 .

- Place the mouse pointer where you want the upper left corner of the rectangular zoom region to be. Depress the move button.

- Drag the mouse pointer to the right and down to complete the zoom rectangle. Release the mouse button.

- Use ⊹⊕⊹ to get an even better approximation of the solution.

To reset the original scale settings:

- Click on **Graph** , **Graph Scaling...** , **Automatic** .

- Click on **OK** .

Graph 1

Graph Title

Y axis

X axis

—X vs. F(X)
+X vs. G(X)

PAGE01: IN US. GRAL

Solving an Equation

You can solve an equation in one variable, for example $\dfrac{3}{2x-1} = \dfrac{x}{3x+1}$, by graphing $f(x) = \dfrac{3}{2x-1}$ and $g(x) = \dfrac{x}{3x+1}$ and then using the trace and zoom features as previously described. There are, however, other ways you can use *f(g) Scholar*™ to solve an equation.

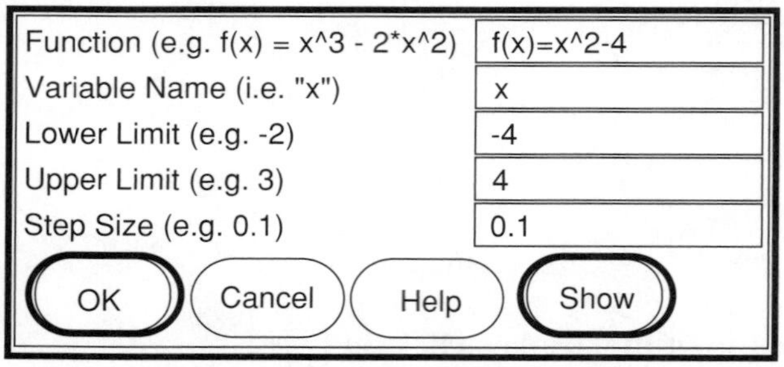

Function (e.g. f(x) = x^3 - 2*x^2)	f(x)=x^2-4
Variable Name (i.e. "x")	x
Lower Limit (e.g. -2)	-4
Upper Limit (e.g. 3)	4
Step Size (e.g. 0.1)	0.1

(OK)   (Cancel)   (Help)   (Show)

To use the numeric solver in *f(g) Scholar*™, follow these steps. Consider the equation $x^2 - 4 = 0$.

- Place the cursor in cell A1 of the spreadsheet.
- From the main menu, select **Calculator** .
- Select the **Math Templates ▶** submenu.
- Select **Roots of a Function. . .** . You will see a dialog box like the one shown here. Enter the information called for.
- Click on **OK** , then **OK** .

The roots will then be displayed. You will be asked if you want the results displayed in the spreadsheet. Click on **Yes** . You will see a spreadsheet display like the one shown.

	A	B	C	D
1	The Roots of f from -4.000000 to 4.000000 are:			
2	Root	Value (x)	Y(x)	
3	1	-2	0	
4	2	2	0	

Spreadsheet 1

Suppose that you wish students to explore zeros of related functions such as these.

$$a(x) = x^2 - 4 \quad b(x) = x^2 - 9 \quad c(x) = x^2 - 16$$
$$d(x) = x^2 - 25 \quad e(x) = x^2 - 36 \quad f(x) = x^2 - 49$$

Follow these steps to find the roots of functions a through f.

- Place the cursor in cell A1 of the spreadsheet.
- Carry out the steps already described to find the roots of $a(x) = x^2 - 4 = 0$. (However, this time, set the lower limit for x at -10 and the upper limit at 10.)
- Place the cursor in cell A5 of the spreadsheet.
- Carry out the steps already described to find the roots of $b(x) = x^2 - 9 = 0$.
- Place the cursor in cell A9. Proceed in the same fashion to find the roots of functions c, d, e, and f.

Spreadsheet Formulas

		A	B	C
Spreadsheet 1				
1	1	13062.5	13187.5
2	2	13650.31	13912.81
3	3	14264.58	14678.02
4	4	14906.48	15485.31
5	5	15577.27	16337
6	6	16278.25	17235.54
7	7	17010.77	18183.49
8	8	17776.26	19183.58
9	9	18576.19	20238.68
10	10	19412.12	21351.81
11	11	20285.66	22526.16
12	12	21198.52	23765.09
13	13	22152.45	25072.17
14	14	23149.31	26451.14
15	15	24191.03	27905.96
16	16	25279.63	29440.78
17	17	26417.21	31060.03
18	18	27605.98	32768.33
19	19	28848.25	34570.59
20	20	30146.43	36471.97
21	21	31503.01	38477.93
22	22	32920.65	40594.21
23	23	34402.08	42826.89
24	24	35950.17	45182.37
25	25	37567.93	47667.4
26	26	39258.49	50289.11
27	27	41025.12	53055.01
28	28	42871.25	55973.04
29	29	44800.46	59051.56
30	30	46816.48	62299.39
31	31	48923.22	65725.86
32	32	51124.76	69340.78

In the *f(g) Scholar*™ spreadsheet, you can enter data, formulas, and text. You can obtain graphical, numerical, and statistical information from it. On this page, you will see how to use the spreadsheet for some financial analysis.

Consider the question below.

If \$12,500 is invested at an annual interest rate of 4.5%, the interest is compounded annually, and the money is left untouched in the bank for n years, how much money A in dollars will be in the account after n years?

You will need: $A = 12,500(1.045)^n$.

To explore the question via the *f(g) Scholar*™ spreadsheet, follow these steps.

- In cell A1, enter 1.

- Select cell A2. Enter A1+1.

- Highlight cell A2 and several cells in column A, for example, A2 through A32.

- From the main menu, select **Edit**, then **Fill**.

- In cell B1, enter the following formula. ROUND(12500*1.045^A1,2)

- Highlight cell B1 through, for example, B32.

- From the main menu, select **Edit**, then **Fill**.

You will see a spreadsheet that shows the amount in the bank (column B) after each year the money is in the bank (column A).

To compare the amount in the account given an interest rate of 5.5% with the amount given at an interest rate of 4.5%, complete column C as follows.

- In cell C1, enter the following formula. ROUND(12500*1.055^A1,2)

- Highlight cell C1 and several cells in column C, for example, C1 through C32.

- Select **Edit**, then **Fill**.

f(g) Scholar™

Recursive Sequences in the Spreadsheet

The spreadsheet can be used to make function tables. If the values of the domain are positive integers, then the function defines a sequence explicitly. There are, however, many problems and questions that require a different type of logic. Consider, for example, the sequence defined by the following recursive formula.

$$\begin{cases} s_1 = 2, \\ s_n = 1.5s_{n-1} & n \geq 2 \end{cases}$$

To use the *f(g) Scholar*™ spreadsheet to track the sequence, proceed as follows.

Spreadsheet 1

	A	B
1	2	3
2	3	4.5
3	4.5	6.75
4	6.75	10.125
5	10.125	15.1875
6	15.1875	22.78125
7	22.78125	34.171875
8	34.171875	51.2578125
9	51.2578125	76.8867187
10	76.8867187	115.330078

- Click on the spreadsheet window to make it active.
- From the main menu, select **Spreadsheet**, then **Options ▶**.
- From the Options submenu, select **Calculation...**.
- Click on **Natural**. Click on **Close**.
- In cell A1, enter 2.
- In cell B1, enter 1.5*A1.
- In cell A2, enter B1.◄—— *The new input is the previous output.*
- Highlight cell A2 and several cells in column A, for example, cells A2 through A10.
- Select **Edit**, then **Fill**.
- Highlight cell B1 and several cells in column B, for example, cells B1 through B10.
- Select **Edit**, then **Fill**.

Consider this sequence. $\begin{cases} s_1 = 1 \\ s_n = \dfrac{1}{1+s_{n-1}} & n \geq 2 \end{cases}$

If you allow n to take on any positive integer value, the sequence will contain infinitely many terms. Perhaps the terms get closer and closer to some definite real number.

To explore this sequence, make the following changes to the procedure already outlined.

- In cell A1, enter 1.
- In cell B1, enter 1/(1+A1).
- In cell A2, enter B1.
- Highlight cell A2 and several cells in column A, for example, cells A2 through A40.
- Select **Edit**, then **Fill**.
- Highlight cell B1 and several cells in column B, for example, cells B1 through B40.
- Select **Edit**, then **Fill**.

Visualizing Spreadsheet Data

In the *f(g) Scholar*™ spreadsheet, you can enter data sets that consist of ordered pairs. Consider this data set.

{(1, 10), (2, 13), (3, 8), (4, 9), (5, 11), (6, 4), (7, 6), (8, 15), (9, 16), (10, 8), (11, 6), (12, 4)}

Enter the first members of the ordered pairs into column A and the second members of the ordered pairs into column B. You can then visualize the data by carrying out the following steps.

- From the main menu, select **Graph**, then **Graph Type ▶**.
- From the Graph Type options, select **XY**.
- Highlight the cells in columns A and B that you wish to graph, in this case A1 through B12.
- From the main menu, select **Spreadsheet**, then **Graph**.

The graph of the data will appear in the graph window.

You may decide that you want to show students the same data presented in a different type of graph.

- From the main menu, select **Windows**, then **New ▶**, then **Graph**.
- From the main menu, select **Graph**, then **Graph Type. . .**.
- From the Graph Type options, select **Vertical Bar**.

To show both graphs on the same display, resize the spreadsheet window and the two graph windows. Then position them side by side as shown.

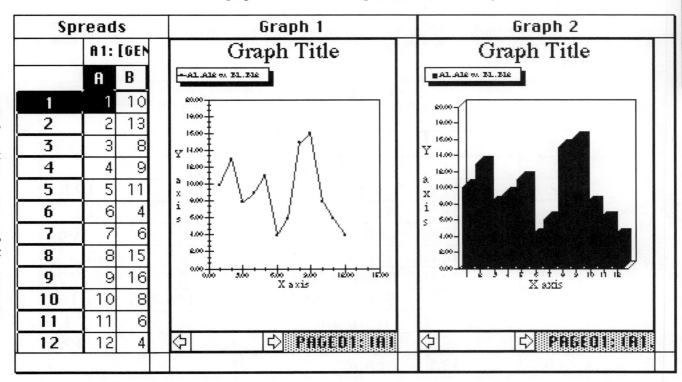

One-Variable Statistical Reports

Consider this data set: {11.6, 11.8, 12.6, 12.1, 12.2, 12.4, 11.8, 11.8, 12.3, 12.5}

A statistical report on the data set contains mean, median, range, and standard deviation.

To create the statistical report in *f(g) Scholar*™, proceed as follows.

- Enter the one-variable data into column A.
- Highlight the range of cells that contains the data.
- From the main tool bar at the top of the computer display, click on .
- From the dialog box, click on **Descriptive**, then select **Statistical Report (Range)**. Click on **OK**.
- The report will appear in columns B and C.

To change the width of one or more columns, highlight those columns. Then:

- Select **Spreadsheet** from the main menu.
- Select **Options ▶**, then **Resize Columns. . .**.
- Enter the column width. (The width shown for columns B and C is 12.) Click on **OK**.

To change the style of the type in one or more cells, highlight those cells. Then:

- Click on **B**, **I**, or **U** for bold, italic, or underlined type, respectively.
- Click on **P**ᴏɪɴᴛ. Then click on the desired type size from the list.
- Click on **F**ᴏɴᴛ. Then click on the desired font from the list. Click on **OK**.

	A	B	C
		Spreadsheet 1	
1	11.6	Sample Size	10
2	11.8	Sum	121.1
3	12.6	Mean	12.11
4	12.1	Pop Var	0.1069
5	12.2	Pop Stdev	0.32695565448544
6	12.4	Sample Var	0.11877777777778
7	11.8	Sample Stdev	0.34464152068168
8	11.8	Minimum	11.6
9	12.3	Maximum	12.6
10	12.5	Range	1

Curve Fitting and Two-Variable Data

Consider the question of fitting an algebraic model to this data set using the spreadsheet.

$$\{(0, 56), (1, 64), (2, 53), (3, 49), (4, 33), (5, 25), (6, 6), (7, -14)\}$$

- Enter the *x*-values of the data into column A and the *y*-values of the data into column B.

- To draw an inference about the appropriateness of a particular model, graph the data in columns A and B in the graph window as an **XY** graph.

- Again highlight the data in columns A and B.

- From the tool bar at the top of the display, click on ⊞. In the resulting dialog box, click on **Polynomial Fit: Y=a+b*X^1+c*X^2+d*X^3. . .** . Click on **OK**.

- Type 2 as the polynomial curve fit order. Click on **OK**.

- Select **Graph Both Data and Equation**. Click on **OK**.

- In the dialog box, enter the information called for. Click on **OK**.

Move the prediction equation on the spreadsheet to the location shown as follows.

- Highlight the cells that contain the equation. Select **Edit**, then **Cut**.

- Highlight a range of cells, such as A10, B10, and C10. Select **Edit**, then **Paste**.

Remove unwanted information as follows.

- Highlight cells D1 through F8. Select **Edit**, then **Delete**.

Place the graph in the spreadsheet by following this step.

- Highlight a block of cells for your graph to occupy. Select **Draw**, then **Graph**.

	A	B	C	D	E	F
1	0	56	59	Graph Title		
2	1	64	58.5238095			
3	2	53	54.7142857			
4	3	49	47.5714285			
5	4	33	37.0952380			
6	5	25	23.2857142			
7	6	6	6.1428571			
8	7	-14	-14.3333333			
9						
10	Y(X)=59+1.1905*X^1+-1		6667*X^2			

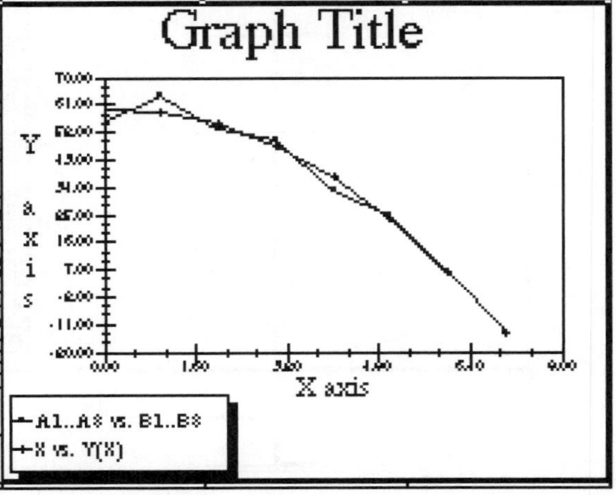

Parametric Equations

The following procedure shows how to graph a curve defined by a pair of parametric equations, such as the curve defined by $\begin{cases} x(t) = 2t + 1 \\ y(t) = -t^2 - 1 \end{cases}$ when $0 \le t \le 4$.

- From the main menu, select **Calculator**, then **Angular Mode ▶**.
- Select **Degrees**.
- Activate the calculator window.
- Type t=0,4,0.1. Press ENTER.
- Type x(t)=2*t+1. Press ENTER.
- Type y(t)=−t^2−1. Press ENTER.
- Type graph(x(t),y(t)). Press ENTER.

An alternative approach to graphing a pair of parametric equations is the following.

- Click on ⎡℞⎤ in the tool bar at the top of the computer display.
- Fill in the information called for.
- Click on **OK**.

If the domain for the variable *t* is a subset of the real numbers and trigonometric functions are involved, begin by selecting radian mode instead of degree mode. The following diagram shows the work needed to graph

$$\begin{cases} x(t) = 2.4 \sin (3.2t - 10) \\ y(t) = 3.3 \cos (4.2t + 2) \end{cases} \text{ when } -2\pi \le t \le 2\pi.$$

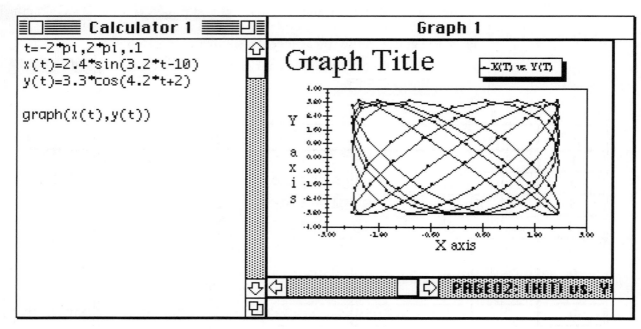

f(g) Scholar™

Creating a Student Handout

In *f(g) Scholar*™, you can create your own handouts. A sample is shown on the next page. You might study it before reading the creation instructions that follow.

To create a student handout like the one on the next page, follow these steps.

- Enter the numerical data into cells A1 through B15.
- Highlight the data. Select ▦ from the tool bar at the top of the display.
- Click on **Polynomial Fit: Y=a+b*X^1+c*X^2+d*X^3. . .** . Click on **OK** .
- Type 2 as the polynomial curve fit order. Click on **OK** .
- Select **Graph Both Data and Equation** . Click on **OK** .

To delete any unwanted statistical information, highlight it. Select **Edit** , then **Delete** .

To move numerical data from cells A1 through B15, highlight those cells. Use the mouse pointer to drag the data to cells A7 through B21.

To edit the text in the graph, double-click on the text you wish to change. A paragraph editor box will appear. You may wish to change the variable names from *x* and *y* to *t* and *h* or change "Graph Title" to "Projectile Curve Fit." After editing, click on the close box of the paragraph editor.

To bring your graph into the spreadsheet, follow these steps.

- Highlight a range of cells in the spreadsheet.
- From the main menu, select **Draw** , then **Graph** .

(Note: Be careful in selecting a region for the graph. It will be made to fit into the space you highlight.)

To type prose into the spreadsheet, follow these steps.

- Highlight a region for your text. Click on 🅣 from the main toolbar.
- Type your prose, for example, questions for students to work on.
- Click on the close box of the paragraph editor.

To change the style of type within an entire block of text, click on the text to select it. Then:

- For bold, italic, or underlined type, click on **B** , **I** , or **U** respectively.
- Click on **Point** or **Font** and click on the desired type size or font from the list that appears. Click on **OK** .

To print your handout after you are satisfied with it, follow these steps.

- Highlight the region that contains your handout or select **Edit** , then **Select All** .
- From the main menu, select **File** , then **Print. . .** . Click on **Print** .

A Study in Projectile Motion

Column A contains a record of the time after launch (t in seconds) of a projectile. Column B records the altitude of the projectile after t seconds (h(t) in feet).

Time	Altitude	h(t) =	28.94 + 130.72*t - 16.10*t^2
0	30	28.94	
0.5	94	90.27	
1	141	143.56	
1.5	183	188.79	
2	226	225.97	
2.5	260	255.1	
3	273	276.18	
3.5	285	289.21	
4	300	294.18	
4.5	291	291.11	
5	283	279.99	
5.5	260	260.82	
6	231	233.59	
6.5	201	198.32	
7	153	154.99	

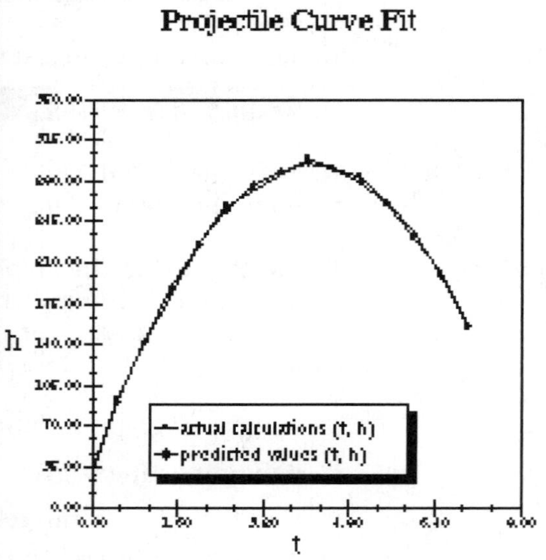

Projectile Curve Fit

h

—▲— actual calculations (t, h)
—◆— predicted values (t, h)

t

Questions for Discussion

1. Refer to columns A and B. Describe the trend in altitude values as the time values increase.

2. Find the difference between the actual altitude and the predicted altitude for t = 4.5 s. Which value is greater?

3. What significance can you give to the ordered pairs (0, 30) and (0, 28.94)?

4. The acceleration due to gravity is 32 ft/s^2. What part of the prediction equation contains this constant?

5. Use any method to find the approximate value of t at which the projectile will strike the ground.

Introduction to Geometry Software in the Classroom

For a long time, people both inside and outside the mathematics classroom viewed the study of geometry as a static enterprise. In that enterprise, students were presented with a set of facts created (definitions), accepted without argument (postulates), and logically arrived at by deduction (theorems). The teacher passed these facts on with some attempt to justify their truth. In turn, students were asked to construct their own proofs to ground the facts presented them and to justify new facts that sprung from them.

The availability of geometry software has changed the enterprise from primarily static to heavily dynamic. The word "dynamic" as used here refers to a study in which virtually thousands of particular instances can be examined with the simple click of a mouse button attached to a computer.

As a result of the computer's capability, students can test various conjectures to discern which are true, which are true only in a restricted context, and which fall down as the exploration continues. From this point of view, students are prompted to think critically and are encouraged not to accept conclusions simply because someone else tells them they are true.

As you know or will see through the use of the following pages, the application of computer technology requires that both teacher and student think about geometric objects, relationships, and operations differently than before. They can now be thought of as tools used to create and conduct geometric experiments. From this point of view, students can effectively learn, or test their understanding of, definitions and facts. For example, suppose that a student wishes to make an enlargement of a geometric object. He or she will need to know the following.

- how to make the initial object

- the fact that the enlargement transformation requires a center and scale factor as well as an object to be enlarged

- the fact that, in a software program, "dilation" is the vocabulary used

- how to take steps in a certain order determined by the software to make the enlargement happen

There are many benefits that result from the use of geometry software. Among them are the following.

- Students learn through an active hands-on methodology.

- The methodology is contemporary, and therefore can generate motivation by its very application.

- The personal student involvement can help generate a sense of responsibility for one's thoughts and work and an accompanying sense of pride and empowerment.

- The experience spawns educational by-products, such as communication skills and social interaction opportunities.

- The mathematical and technological lessons learned will help students participate more confidently and competently in the adult world.

The pages that follow contain activities designed for each of the following geometric software programs: *The Geometer's Sketchpad®* (version 3.0), *Cabri Geometry II™* (version 1.0), and *The Geometric superSupposer*. Since the activities are different for each software program, you may decide to read some of the activities written for other software programs and adapt them for use with your software.

The Geometer's Sketchpad® Basics

A diagram that illustrates some of the basic components of *The Geometer's Sketchpad®* is shown on this page. The diagram illustrates the following.

- drawing space
- drawing tools
- selection tool
- text tool
- document title
- scroll controls that allow you to move across and down in the drawing space
- controls that enable you to enlarge or reduce the visible portion of the drawing space
- the close-document box

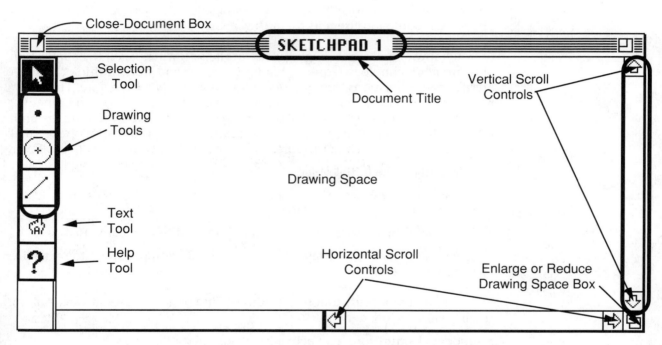

- To open, or launch, *The Geometer's Sketchpad®*, highlight the application name or icon or highlight the name or icon of a particular drawing document. Then double-click on it.
- To close a displayed document, click on the small square at the upper left corner of the document.
- To end the session by closing *The Geometer's Sketchpad®* and all open documents, click on **File**, then **Quit**.
- To open a document when the application has already been launched, click on **File**, then **Open...**. In the list that appears, select the name of the document you wish to use.
- To save a document you have created, select **File**, then **Save As...**. Type a name for your document and click on **Save**.

The Geometer's Sketchpad® Menus and Tools

The Geometer's Sketchpad® (version 3.0) is easy to use because many of its capabilities are grouped in menus whose titles you will see at the top of the computer display. As you can see from the menus shown here, a simple click of the mouse button will give you access to a large collection of tools that you can use separately or together.

Note: When you explore the menus, you may find that some of the items listed are different from those shown here. The differences are due to previously chosen selections in the toolbar at the left of the computer display or to an action taken previously.

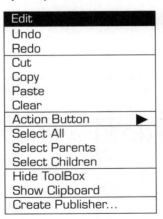

Construct
Point On Object
Point At Intersection
Point At Midpoint
Line
Perpendicular Line
Parallel Line
Angle Bisector
Circle By Center+Point
Circle By Center+Radius
Arc On Circle
Arc Through 3 Points
Interior
Locus
Construction Help...

Transform
Translate...
Rotate...
Dilate...
Reflect
Mark Center
Mark Mirror
Mark Vector
Mark Distance
Mark Angle
Mark Ratio
Define Transform...

Measure
Distance
Length
Slope
Radius
Circumference
Area
Perimeter
Angle
Arc Angle
Arc Length
Ratio
Coordinates
Equation
Calculate...
Tabulate
Add Entry

Also found in the main menu are groups of tools that you can use to edit and stylize your work and a menu that allows you to work in a coordinate plane.

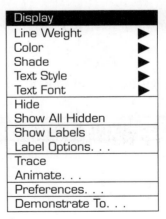

Edit
Undo
Redo
Cut
Copy
Paste
Clear
Action Button ▶
Select All
Select Parents
Select Children
Hide ToolBox
Show Clipboard
Create Publisher...

Display
Line Weight ▶
Color ▶
Shade ▶
Text Style ▶
Text Font ▶
Hide
Show All Hidden
Show Labels
Label Options...
Trace
Animate...
Preferences...
Demonstrate To...

Graph
Create Axes
Show Grid
Snap To Grid
Grid Form ▶
Plot Measurement...
Plot Points...
Coordinate Form ▶
Equation Form ▶

On the pages that follow, you will see how to use the commands in these menus to set up classroom demonstrations and classroom explorations. You will also see how to use these toolbar icons that appear to the left of the drawing space.

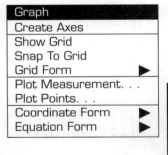

GEOMETRY SOFTWARE

Polygons and Measurements

Fundamental to any study of geometry is the creation of geometric figures and measurements of their parts. What follows is a simple procedure for creating a triangle in *The Geometer's Sketchpad®* and measuring its sides and angles.

- Using the mouse pointer, select the point tool, .

- In the drawing space, click once for the first vertex of the triangle.

- Place the mouse pointer on a second location in the drawing space. Click once.

- Place the mouse pointer on a third location in the drawing space. Click once.

- With | SHIFT | depressed, click on each of the three points to select them.

- From the main menu, select **Construct** , then **Segment** .

You have now created a triangle. To change the triangle that you made:

- Select [▶] . Place the mouse pointer on a vertex.

- Press the mouse button. Keep it pressed as you drag that vertex to another location.

- Then release the mouse button.

Notice that the triangle shown is labeled. You may wish to change the labels to others. For example, to change the label A to the label Z, proceed as follows.

- Select [▶] . Using the mouse pointer, click on the point labeled A.

- From the main menu, select **Display** , then **Relabel Point...** .

- Type Z in the highlighted space in the dialog box that appears. Click on **OK** .

You will need to measure sides and angles. Here is how to find the length of $\overline{BC}$.

- Select [▶] . Using the mouse pointer, click on $\overline{BC}$, but not at an endpoint.

- From the main menu, select **Measure** , then **Length** .

The measure of $\overline{BC}$ will appear on the screen. To move the measurement to another screen location, use the mouse pointer and drag it.

To measure $\angle ABC$, follow these steps.

- Select [▶] . Place the mouse pointer on point A.

- With | SHIFT | depressed, place the mouse pointer on point B and click once, then on point C and click once.

- From the main menu, select **Measure** , then **Angle** .

SKETCHPAD 1

k = 2.190 inches
m∠ABC = 85.06°

Select Measurement m1

Understanding Definitions

An interesting and probing activity involving geometric definitions can be made in *The Geometer's Sketchpad®* by constructing a quadrilateral and measuring its parts. After constructing a diagram like the one shown, consider different ways to deform the quadrilateral so that it becomes a special quadrilateral, such as a rhombus, trapezoid, or kite. To create a diagram like the one shown:

- Select the point tool from the toolbar at the left of the sketchpad. Using the procedure described at the top of page 124, create a quadrilateral.

- To display the measures of the sides, hold down SHIFT and click on each of the sides to select them. From the main menu, select **Measure**, then **Length**. All four lengths will appear at once.

- To display the measure of an angle, click on three vertices while holding down SHIFT. From the main menu, select **Measure**, then **Angle**.

- Repeat this process three more times. (The angle whose measure you get is the angle whose vertex is the second point you chose in your selection of three vertices.)

Now think about how to deform quadrilateral *ABCD* so that it represents a parallelogram. For example, you know that, by definition, a parallelogram is a quadrilateral in which both pairs of opposite sides are parallel. Ask yourself how you can use the displayed measures to adjust *ABCD* so that it satisfies this requirement.

You might also want to explore the properties of the diagonals of a particular type of quadrilateral.

- Holding down SHIFT, click on two opposite vertices. From the main menu, select **Construct**, then **Segment**.

- Repeat this process with the other pair of opposite vertices.

- Holding down SHIFT, click on the two diagonals with the mouse pointer.

- From the main menu, select **Construct**, then **Point At Intersection**.

Using the modified diagram, ask yourself what must be true of the diagonals of a quadrilateral if it is an isosceles trapezoid. Make the isosceles trapezoid. Then use the measurement tools to check the conjecture. You should find that the diagonals form four smaller segments that are congruent in pairs.

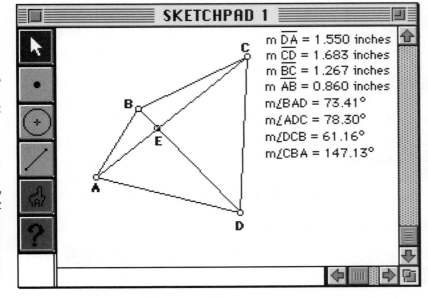

SKETCHPAD 1

m $\overline{DA}$ = 1.550 inches
m $\overline{CD}$ = 1.683 inches
m $\overline{BC}$ = 1.267 inches
m $\overline{AB}$ = 0.860 inches
m∠BAD = 73.41°
m∠ADC = 78.30°
m∠DCB = 61.16°
m∠CBA = 147.13°

GEOMETRY SOFTWARE

Parallel and Perpendicular Lines

The Geometer's Sketchpad® has the capability of constructing parallel and perpendicular lines. What follows will show you how to construct a rectangle by using the parallel and perpendicular line commands.

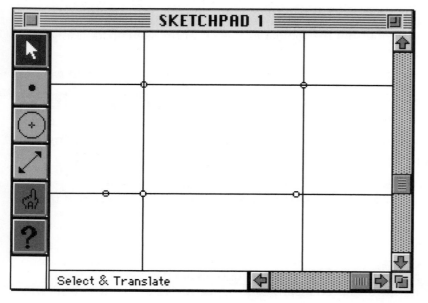

- Press the spacebar until the line tool, , appears in the toolbar to the left of the drawing space. Then select the line tool.

- In the drawing space, draw a line. Position it so that it is horizontal.

- From the main menu, select **Construct**, then **Point On Object**.

- Select . With SHIFT depressed, click on the line.

- From the main menu, select **Construct**, then **Perpendicular Line**.

- Using both **Construct**, and **Point On Object**, place a point on the vertical line you just made.

- With SHIFT depressed, click on the line just made. From the main menu, select **Construct**, then **Perpendicular Line**.

- Proceed in this fashion to construct the fourth side of the rectangle.

The construction just described is based on using the perpendicular line command three times. You could construct a rectangle by using the perpendicular line command twice and the parallel line command once.

- Select the line tool, .

- In the drawing space, draw a line. Position it so that it is horizontal.

- Select the point tool, . Click on a location not on the line.

- Select . With SHIFT depressed, click on the line you made.

- From the main menu, select **Construct**, then **Parallel Line**.

- From the main menu, select **Construct**, then **Point On Object**.

- With both the point and the line just made selected, select **Construct**, then **Perpendicular Line**.

- Proceed in the same way to construct the fourth side of the rectangle.

Ask yourself what other ways you can construct a rectangle. Explore those possibilities.

GEOMETRY SOFTWARE

Reflections, Rotations, and Translations

You will probably want to introduce your geometry students to transformations such as reflections and rotations. *The Geometer's Sketchpad®* has both these transformations built into it. They can be accessed by choosing **Transform**.

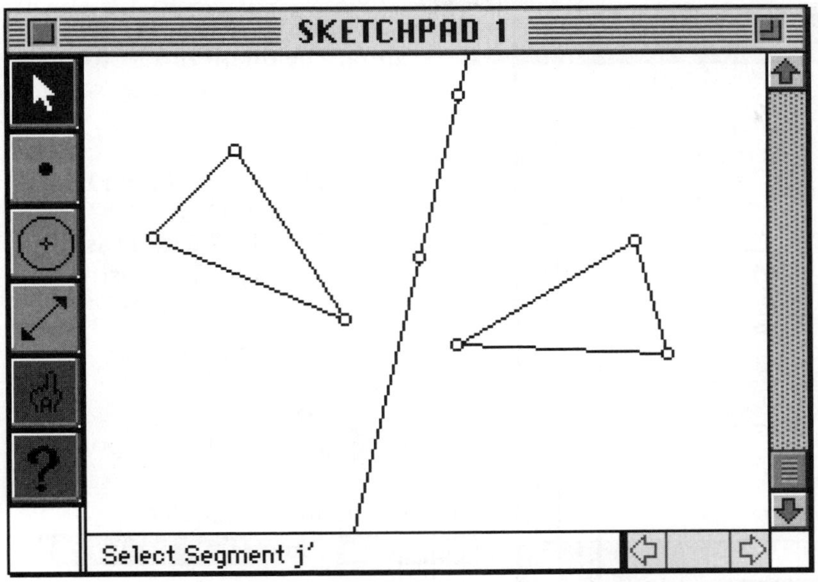

For example, suppose you want to reflect a triangle across a line. Begin by drawing the triangle and line as previously described. Then follow these steps.

- Select **⬆**. Click on the line.

- From the main menu, select **Transform**, then **Mark Mirror "m"**. (The label for your line may be different from *m*.)

- With **SHIFT** depressed, click on each of the vertices and sides of the triangle.

- From the main menu, select **Transform**, then **Reflect**.

You should now see the original triangle, the original line, and the reflection image of the original triangle, as shown here.

The steps involved in rotating a geometric figure about a point are somewhat similar to those needed to reflect a figure across a line. To experiment with the process, select the line of reflection that you drew. Delete it and any points on it. (To delete items, select them. Choose **Edit**, then **Clear**.) Using the point tool, place a point in the drawing space. Then follow these steps.

- Select **⬆**. Click on the point.

- From the main menu, select **Transform**, then **Mark Center "F"**. (The label may be different from *F*.)

- With **SHIFT** depressed, click on each of the vertices and sides of the triangle.

- From the main menu, select **Transform**, then **Rotate...**.
  In the dialog box that appears, enter the number of degrees of the rotation. (A positive number will result in a counterclockwise rotation. A negative number will result in a clockwise rotation.) Click on **OK**.

A translation is the easiest transformation to perform. With **SHIFT** depressed, click on each of the vertices and sides of the triangle. From the main menu, select **Edit**, then **Copy**, then **Edit**, then **Paste**. Select **⬆**. Place the mouse pointer on any side of the copy, depress the mouse button, and drag the copy to the desired location.

SKETCHPAD 1

Select Segment j'

Dilations and Similarity

Reflections, rotations, and translations are particular types of geometric transformations. A dilation is another type. Basically, the dilation is the scaling up or down of a geometric figure with a particular point chosen as the center of the scaling. The dilation transformation is built into *The Geometer's Sketchpad*®.

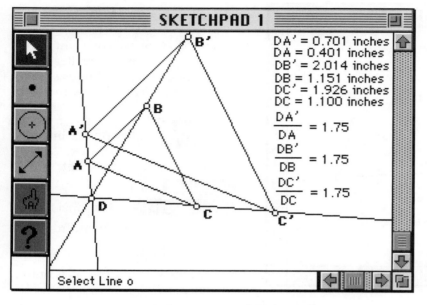

The sketchpad diagram shown here represents the dilation of $\triangle ABC$ scaling by a factor of 1.75 with D as the center of the dilation. To construct this drawing, take these steps.

- Draw $\triangle ABC$ and point D as previously described.

- Select [ ] and click on point D.

From the main menu, select **Transform**, then **Mark Center "D"**.

- Select [ ]. Click on point A. With [SHIFT] depressed, click on all the other vertices and sides of $\triangle ABC$.

- From the main menu, select **Transform**, then **Dilate...**. Type 1.75. Click on **OK**.

- Select the line tool, [ ]. Select [ ] and click on point D. With [SHIFT] depressed, click on A'. From the main menu, select **Construct**, then **Line**.

- Proceed in the same fashion to get lines through D and B' and D and C'.

To illustrate the similarity between $\triangle A'B'C'$ and $\triangle ABC$:

- From the main menu, select **Measure**, then **Distance** after choosing D and A', D and A, D and B', D and B, D and C', and D and C, respectively.

- From the main menu, select **Measure**, then **Calculate...**. In the drawing space, highlight DA' = ... In the calculator, click on **/**. In the drawing space, highlight DA = ... Click on **OK**. Proceed in similar fashion to display the other two similarity ratios.

You can also construct multiple dilations of a given figure. For example, start with $\triangle ABC$ and point D. Construct $\triangle A'B'C'$ similar to $\triangle ABC$ with scale factor 1.5 as above. Then construct $\triangle A''B''C''$ similar to $\triangle ABC$ with scale factor 2.

Is it true that $\triangle A'B'C'$ is similar to $\triangle A''B''C''$? If so, what is the similarity ratio? You should be able to demonstrate that any pair of dilations of a given figure are similar to one another.

Dynamic Drawings and Conjectures

The Geometer's Sketchpad® diagram shown can be used to help students conclude that if one vertex of a triangle is moved along a line parallel to the opposite side,

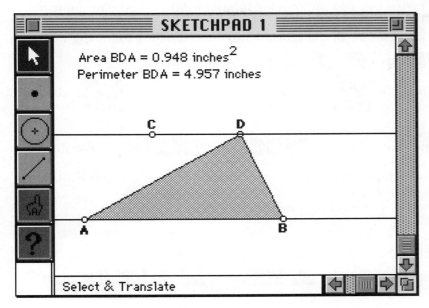

then the area of the triangle remains constant, but the perimeter does not.

To set up the exploration, you will need to sketch a line $\overleftrightarrow{CD}$ that not only is parallel to $\overline{AB}$, but also does not move when point D is dragged along it. To create such a dynamic figure, proceed as follows.

- Using the line tool, sketch a horizontal line $\overleftrightarrow{AB}$.

- Using the point tool, locate a point C not on $\overleftrightarrow{AB}$.

- Select [pointer] and click on point C.

  With ⎡SHIFT⎤ depressed, click on $\overleftrightarrow{AB}$. From the main menu, select **Construct**, then **Parallel Line**.

- From the main menu, select **Construct**, then **Point On Object** . (In the diagram on this page, that point is labeled D.)

- **Draw** $\overline{AD}$ and $\overline{DB}$ as previously described.

- Select [pointer] and click on point A. With ⎡SHIFT⎤ depressed, click on points B and D. From the main menu, select **Construct**, then **Polygon Interior** .

- Click the mouse pointer in the triangle's interior and select **Measure**, then **Area** from the main menu.

- Click the mouse pointer in the triangle's interior and select **Measure**, then **Perimeter** from the main menu.

As you drag point D along $\overleftrightarrow{CD}$, you should see that the area does not change, but the perimeter does.

With this dynamic diagram, you can search for that position of point D on $\overleftrightarrow{CD}$ that gives the minimum perimeter of $\triangle ABD$.

Suppose that $\overleftrightarrow{CD}$ is moved farther away from $\overleftrightarrow{AB}$. You will find that the minimum perimeter changes. [It increases.] However, you will also find that the minimum still occurs when $\triangle ABD$ is isosceles.

Then you might explore why it is that the area of $\triangle ABD$ does not change when point D is dragged along $\overleftrightarrow{CD}$ as described. [Base and height remain constant.]

Theorems and the Coordinate Approach

In modern geometry education, students are exposed to many conceptual approaches. *The Geometer's Sketchpad®* not only can help you with topics from synthetic geometry, but also can provide support for the coordinate approach.

The Geometer's Sketchpad® diagram below was created by the following steps.

- Select **Graph**, then **Create Axes** from the main menu.

- Using the line tool, [line tool icon], draw a line. Using the text tool, [text tool icon], label the line $\overleftrightarrow{EF}$.

- With the line selected, choose **Measure**, then **Equation** from the main menu.

With this construction as a basis, you can extend it to help students explore geometric questions from a coordinate point of view. For example, the following steps illustrate how you might explore this theorem.

Two nonvertical lines are perpendicular if and only if the product of their slopes is –1.

- Select $\overleftrightarrow{EF}$, the line you just drew.

- From the main menu, select **Construct**, then **Point On Object**. Label the new point G.

- With both $\overleftrightarrow{EF}$ and point G selected, select **Construct**, then **Perpendicular Line**.

- With the new line selected, select **Construct**, then **Point On Object**. Label the new point H.

- With $\overleftrightarrow{GH}$ selected, select **Measure**, then **Equation**.

- With $\overleftrightarrow{EF}$ selected, select **Measure**, then **Slope**.

- With $\overleftrightarrow{GH}$ selected, select **Measure**, then **Slope**.

- From the main menu, select **Measure**, then **Calculate...**.

- In the drawing space, highlight the slope of $\overleftrightarrow{EF}$. In the calculator, click on **∗**.

- In the drawing space, highlight the slope of $\overleftrightarrow{GH}$. In the calculator, click on **OK**.

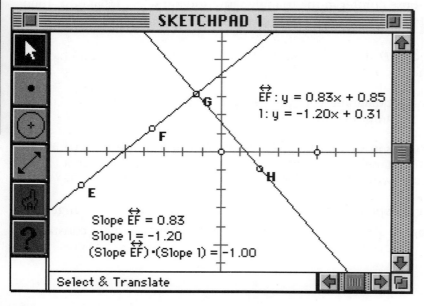

You should see a display like the one shown here.

If you now select point E, F, or H and drag it, you will see that the lines remain perpendicular and the product of their slopes remains constant at –1.

There are many ways to extend this exploration. For example, add $\overleftrightarrow{EH}$ to the drawing. Compute and display its slope. Then compute and display the product of the slopes of $\overleftrightarrow{EH}$ and $\overleftrightarrow{EF}$. Notice that the product is not equal to –1. Thus, the lines are not perpendicular. This gives another basis for the statement that a triangle cannot have two right angles.

Intersecting Lines and Linear Systems

A system of two linear equations in two unknowns is shown here.

$$\begin{cases} y = 1.20x + 0.29 \\ y = -0.92x - 0.49 \end{cases}$$

You may be surprised to learn that you can use *The Geometer's Sketchpad*® to solve it. The following is a strategy for doing so.

- From the main menu, select **Graph**, then **Create Axes**.
- From the main menu, select **Graph**, then **Equation Form ▶**, then **Slope/Intercept**. (This option may already have a check mark next to it. If so, you may simply release the mouse button.)
- Using the line tool in the toolbar at the left of the drawing space, construct two lines that intersect.
- Select both lines by clicking on them while holding down | SHIFT |.
- Select **Measure**, then **Equation** from the main menu. You will now see an equation in slope-intercept form for each equation in the system.
- Rotate the lines by dragging a point on each until the slopes of the lines agree with the slopes in the given equations above.
- Translate the lines by clicking on each and dragging it until the *y*-intercept agrees with the *y*-intercept in the appropriate equation.
- Holding down | SHIFT |, click on each of the two lines. From the main menu, select **Construct**, then **Point At Intersection**.
- Click on the point of intersection. From the main menu, select **Measure**, then **Coordinates**.

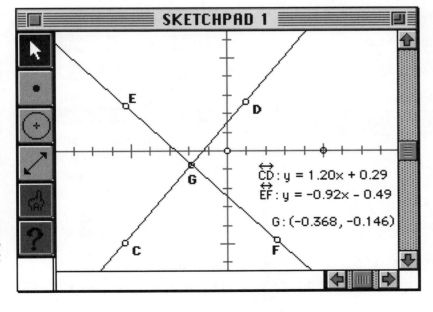

The coordinates just found approximate the solution of the given linear system. For this reason, the coordinates on your screen may differ slightly from those shown here.

Suppose you wish to solve a system like the following.

$$\begin{cases} 2.5x - 3y = 13 \\ 1.5x + 1.5y = 10 \end{cases}$$

Follow the same procedure but with this change in the second step.

- From the main menu, select **Graph**, then **Equation Form ▶**, then **Standard**.

The Unit Circle and Trigonometric Ratios

To set up a demonstration of trigonometric ratios, you will need a circle of radius 1 centered at the origin of a rectangular coordinate system. You will also need to construct points and lines that determine a right triangle, $\triangle FIJ$, like the one in this diagram. Proceed as follows.

- From the main menu, select **Graph**, then **Create Axes**.

- Using the circle tool, , draw a circle. Use the text tool to label the center F and the point on the circle G.

- Click on the circle and drag it until point F is at the origin.

- Click on the circle, then select **Measure**, then **Equation**.

- Click on point G and drag it until the radius is 1.

- With the circle selected, select **Construct**, then **Point On Object**. Label the new point I.

- Select points F and I while holding down SHIFT . Select **Construct**, then **Line**.

- Click on the horizontal axis and point I while holding down SHIFT .

- Select **Construct**, then **Perpendicular Line**.

- Select the line just drawn and the horizontal axis. Select **Construct**, then **Point At Intersection**. Label the new point J.

If you now drag point I around the circle, you will see how $\triangle FIJ$ changes.

To explore the trigonometric ratios of $\angle IFJ$, proceed as follows.

- Select F and J holding down SHIFT . Select **Measure**, then **Distance**.

- Select I and J holding down SHIFT . Select **Measure**, then **Distance**.

- Select F and I holding down SHIFT . Select **Measure**, then **Distance**.

- From the main menu, select **Measure**, then **Calculate...**.

- In the drawing space, highlight IJ = ... In the calculator, click on **/** . In the drawing space, highlight FI =.... In the calculator, click on **OK** . (This ratio is sin F.)

- Follow the same procedure to calculate the ratios $\frac{FJ}{FI}$ and $\frac{IJ}{FJ}$, the cosine and tangent of $\angle F$, respectively.

Enlarge or reduce the circle by dragging point G. You will see that, for a given angle, the three trigonometric ratios remain constant.

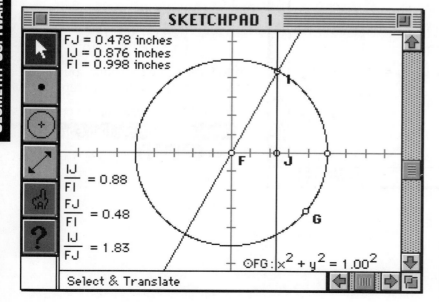

Cabri Geometry II™ Basics

As you can see from this diagram of the *Cabri Geometry II™* drawing sheet, *Cabri Geometry II™* is an icon-based geometry software program. That is, there are few items in the main menu, **File**, **Edit**, and **Options**. The tools you will need are found above the drawing sheet in the toolbar.

What follows is a brief tour of the tools available to you. The actual icon you see may be different if one of the tools associated with the icon shown on this page was recently chosen and saved prior to the time you opened the application.

File   Edit   Options

Icon	Tools available in conjunction with it
	Pointer, Rotate, Dilate, Rotate and Dilate
	Point, Point on Object, Intersection Point
	Line, Segment, Ray, Vector, Triangle, Polygon, Regular Polygon
	Circle, Arc, Conic
	Perpendicular Line, Parallel Line, Midpoint, Perpendicular Bisector, Angle Bisector, Vector Sum, Compass, Measurement Transfer, Locus, Redefine Object
	Translation, Rotation, Dilation, Reflection, Symmetry, Inverse
	Initial Object, Final Object, Define Macro
	Collinear, Parallel, Perpendicular, Equidistant, Member
	Distance & Length, Area, Slope, Angle, Equation & Coordinates, Calculate, Tabulate
	Label, Comments, Numerical Edit, Mark Angle, Fix / Free, Trace On/Off, Animation, Multiple Animation
	Hide/Show, Color, Fill, Thick, Dotted, Modify Appearance, Show Axes, New Axes, Define Grid

Polygons and Measurement

On this page, you will see how polygons are drawn in *Cabri Geometry II*™ and how their sides and angles are measured and displayed.

To create a polygon, follow these steps.

- Click on the icon third from the left in the toolbar. Select ⬟ **Polygon**. The icon will change to 🔲.

- On the drawing pad, click on one location for the first vertex.

- Place the pencil pointer on a second location and click for the second vertex. In similar fashion, create the other vertices.

- For the final side, place the pencil pointer on the first vertex. Click once.

To label the vertices of the polygon, proceed as follows.

- Click on the icon second from the right in the toolbar. Select **Label**. The icon becomes 🔲. Click on a vertex. Type the label name.

- Click on each of the other vertices, typing a label name for each vertex.

To display the length of each of the sides of the polygon, follow these steps.

- From the toolbar, click on the third icon from the right. Select **Distance & Length**. You will see 🔲 appear.

- Click on one endpoint of the side.

- Click on the other endpoint of the side.

- Repeat this procedure for each of the other sides of the polygon.

To display the measure of each of the angles of the polygon, do the following.

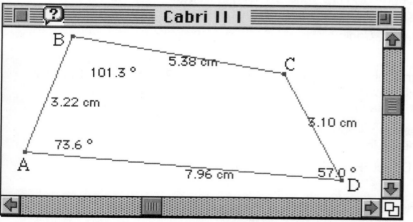

- From the toolbar, click on the icon third from the right. Select **Angle**. You will see 🔲 appear.

- Click on one vertex of the polygon.

- Click on a second vertex of the polygon. (This is the angle whose measure you will see.)

- Click on a third vertex of the polygon. Repeat this procedure for each other angle.

You can now pose and find answers to various questions about the polygon. For example:

- Does it happen to have any special properties?

- Is it convex or concave? How would you change the figure to make it the other possibility?

- How would you alter the polygon so that it is no longer a polygon?

Special Quadrilaterals and the Coordinate Plane

This diagram shows quadrilateral *OABC* on a coordinate plan. With this diagram, you can explore questions dealing with properties of quadrilaterals.

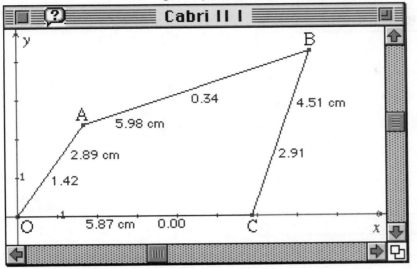

To create the diagram:

- Select the icon at the far right in the toolbar. Select **New Axes** . The icon will become 🔲 .

- Click once to place the origin of the coordinate system.

- Drag the pencil pointer to make a horizontal axis. Click once.

- Drag the pencil pointer until the second axis that shows is vertical. Click once.

- Select the icon third from the left in the toolbar. Select **Segment** . The icon will become 🔲 . Click once on the origin. (This will be labeled O.) Click once on a location in the first quadrant. (This point will be labeled A.)

- Proceed in this manner to make the quadrilateral shown in the diagram.

To label the vertices of the quadrilateral, proceed as follows.

- Click on the second icon from the right in the toolbar. The icon will become 🔲 .

- Click on the origin. Type the label O. Click on the next vertex and type the label A.

- Proceed in the same fashion to label the other vertices.

To display the length and slope of each of the sides of the polygon, do this.

- Click on the icon third from the right in the toolbar. Select **Distance & Length** . You will see 🔲 appear. Click on a side of the quadrilateral. The length will then be displayed. Repeat this process three more times.

- Click on the icon third from the right in the toolbar. Select **Slope** . The icon will become 🔲 . Click on a side of the quadrilateral. The slope will then be displayed. Repeat this process for the other sides.

Consider exploring questions like these.

- If you deform quadrilateral *OABC* until the lengths of opposite sides are equal, what can you expect to be the relationship between slopes of opposite sides?

- Suppose $\overline{OB}$ and $\overline{AC}$ along with their intersection *D* are added to the diagram. What will be true of the four segments that make up the diagonals if quadrilateral *OABC* is deformed to make a rhombus?

Reflections and Rotations

Among the transformations built into *Cabri Geometry II*™ are reflection and rotation. Suppose that you want to reflect pentagon *ABCDE* across a line.

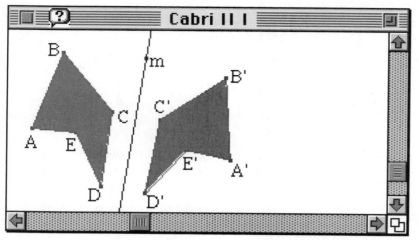

To reflect pentagon *ABCDE* across a line, follow these steps.

- Select the icon that is third from the left in the toolbar. Select **⬠ Polygon**. The icon will become [icon].

- Draw the polygon, in this case a pentagon, as previously described.

- Select the icon second from the right in the toolbar. Select **Label**. Its icon is [AI]. Label the vertices as previously described.

- Select the icon that is third from the left in the toolbar. Select **Line**. The icon will become [icon]. Draw the line that will become the line of reflection.

- Select the icon sixth from the left in the toolbar. Select **Reflection**. Its icon is [icon].

- Click on pentagon *ABCDE*. Click on the line of reflection, which in this diagram is *m*. The reflection image will then appear.

- Label the vertices of the image as previously described.

To rotate pentagon *ABCDE* through an angle such as 45°, follow these steps. (First delete *m* and any points on it.)

- Click on the icon second from the left in the toolbar. Select **Point**. The icon will become [icon].

- Click once to create a point that will serve as the center of rotation.

- Select the icon second from the right in the toolbar and select **Numerical Edit**. Its icon is [2.1]. Click in an open space and type 45.

- Click on the icon sixth from the left in the toolbar. Select **Rotation**. You will see [icon] appear. Click on a side of the pentagon, then on 45, then on the center of rotation.

Consider exploring questions like these.

- Look for conditions under which a polygon and its image under a reflection overlap.

- Suppose you rotate a polygon 180° around one of its vertices. Look for conditions under which the image is also a reflection of the original polygon across a line.

HRW material copyrighted under notice appearing earlier in this work.

Dilations and Similarity

The dilation transformation, also known as a scaling, or the enlargement/reduction transformation, is built into *Cabri Geometry II™*. You perform this transformation in the same way you perform a rotation.

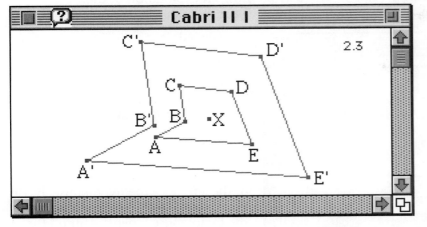

To dilate pentagon *ABCDE* with a scale factor such as 2.3 and center of dilation *X*, follow these steps.

- Sketch and label a pentagon as previously described. (*Note:* In this activity, use the polygon tool rather than the segment tool to construct the polygon.)

- Click on the icon second from the left in the toolbar. Select **Point**.

The icon will become [image]. Click on a location that will serve as the center of the dilation, point X.

- Select the icon second from the right in the toolbar. Then select **Numerical Edit**. Its icon is **2.1**. Click in a convevient location and type 2.3.

- Click on the icon sixth from the left in the toolbar. Select **Dilation**. You will see [image] appear. Click on a side of the pentagon, then on 2.3, then on the center of the dilation.

The computer display will then show the original pentagon, the center of the dilation, and the image under the dilation. Use the steps previously described to label the vertices of the image.

Consider exploring issues like these.

- The diagram on this page shows two polygons related by a dilation with a scale factor of 2.3. Investigate the relationship between the perimeters of the image and preimage. Also investigate the relationship between their areas. (*Note:* If you draw a polygon using the polygon tool and choose the distance and length tool, the perimeter of the polygon will be displayed when you place the mouse pointer on a side of the polygon.)

- Explore how the area of the region between the image and the preimage is related to the area of the image and the area of the preimage. You might begin this exploration by constructing a right triangle, dilating it by a simple scale factor such as 2, and then looking for a pattern. After exploring some simpler geometric figures, test conjectures using more complicated figures such as pentagons.

- Investigate what happens when the center of the dilation is moved from one location to another. For example, consider what happens when the center of the dilation is to the upper left of the computer screen and outside the polygon, when it is at the upper right of the screen and outside the polygon, and so on. Also explore what happens when the scale factor is negative or is a real number between 0 and 1.

The Calculator and the "Pythagorean" Right-Triangle Theorem

The diagram shown can be very useful during an exploration of the "Pythagorean" Right-Triangle Theorem. The display shows

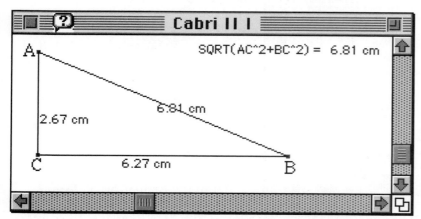

- a labeled right triangle,
- the length of its sides, and
- a calculation of the length of the hypotenuse made in the *Cabri Geometry II™* calculator.

When the right triangle is changed to another one, the lengths of the sides will change and so will the calculation of the length of the hypotenuse. It will then become clear how the length of the hypotenuse is dependent on the lengths of the legs of the right triangle.

To set up the display, follow these steps.

- Sketch and label a right triangle using the segment tool.
- Select the sides and display their lengths as previously described.
- Select the icon third from the right in the toolbar. Then select **Calculate**.

  Its icon is [icon]. At the bottom of the computer screen, you will see the calculator strip.

To calculate *AB*, follow this procedure.

- In the calculator, click on $\boxed{\sqrt{}}$. In the drawing, click on the measurement for the length of $\overline{AC}$.
- In the calculator, click on $\boxed{\wedge}$. Type 2. Click on $\boxed{+}$.
- In the drawing, click on the measurement for the length of $\overline{BC}$.
- In the calculator, click on $\boxed{\wedge}$. Type 2. Click on $\boxed{)}$, then $\boxed{=}$. The result will appear at the right in the calculator strip.

To transfer and edit the calculation:

- Place the mouse pointer on the calculation. Depress the mouse button.
- In the drawing, drag the calculation to a convenient screen location.
- Release the mouse button. Then click once.
- Then type the text that is shown at the upper right in this diagram.

Consider this open-ended extension based on the "Pythagorean" Theorem.

- Add the altitude from vertex *C*. Select the icon fifth from the left. Select **Perpendicular Line**. The icon will become . Click on $\overline{AB}$, then on *C*.

Mark the intersection of the line just drawn and $\overline{AC}$. Explore relationships between the three triangles and their parts.

Bisectors and Triangles

There are many sets of lines that can be drawn in a triangle. These sets include perpendicular bisectors, angle bisectors, medians, and altitudes. A study of any one set of lines in a triangle can prove to be very informative and instructive. This Cabri diagram illustrates the construction of the three perpendicular bisectors and the three angle bisectors in $\triangle ABC$.

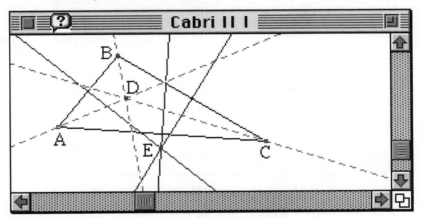

The construction is easy to make and proceeds along these lines.

- Use the segment tool to sketch a triangle, and label it as previously described.

- Select the icon fifth from the left in the toolbar. Select **Perpendicular Bisector**. Its icon is [icon]. Click on each of the three sides of the triangle.

- Select the icon second from the left in the toolbar. Then select **Intersection Point**. Its icon is [icon]. Click on two of the perpendicular bisectors. Label the intersection point. In this diagram that point is E.

- Select the icon fifth from the left in the toolbar. Then select **Angle Bisector**. Its icon is [icon]. Click on three vertices of the triangle. This will give the angle bisector of the angle whose vertex is the second point in the set. Repeat this process two more times.

- Select the icon second from the left in the toolbar. Then select **Intersection Point**. Its icon is [icon]. Click on two of the angle bisectors. Label the intersection point. In this diagram that point is D.

Once the construction has been made, drag one vertex of the triangle to another location. The display will show that no matter what triangle is formed, the perpendicular bisectors are concurrent in a point and so are the angle bisectors.

Consider this related problem.

- Construct the three medians in $\triangle ABC$. Once drawn it will be clear that they too are concurrent.

- Construct the three altitudes in $\triangle ABC$. These three lines are also concurrent.

- Suppose that you construct all four types of lines discussed on this page. Explore the question of how many lines can be put through the set of the four points of concurrency. It may help to notice that the perpendicular bisectors, medians, and altitudes have points of concurrency that are collinear.

Slopes, Lengths, Parallelograms, and Coordinates

The Cabri diagram shows quadrilateral *ABCD* on a coordinate system. How can the quadrilateral be made into a parallelogram? Two methods are shown here. Begin by constructing new axes and a quadrilateral using the line tool.

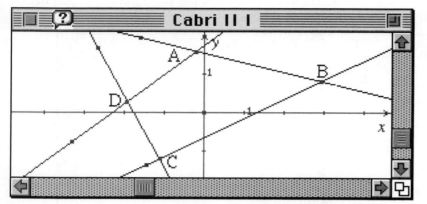

Method 1

- Select the icon third from the right in the toolbar, then **Equation & Coordinates**. Its icon is . Click on a line, and then one of the axes. An equation for that line will appear on the display.

- Repeat this step for each of the other lines that bound the quadrilateral.

- Grab a line and rotate it until its slope is equal to that of the line "opposite" it.

- Repeat this for the other pair of lines bounding the quadrilateral.

If the slopes of two lines are equal, they are parallel. Since this condition is satisfied, quadrilateral *ABCD* is a parallelogram.

Method 2

- Select the icon third from the right in the toolbar. Then select **Distance & Length**. Its icon is `cm↗`. Click on each pair of adjacent vertices to display the lengths of the sides.

- Click on the sides and rotate them until the lengths of opposite sides are equal.

If opposite sides of a quadrilateral are congruent, then it is a parallelogram.

How can the parallelogram be made into a rectangle? Here is one strategy.

- Display the equations of the four lines as described already.

- Select the icon third from the right in the toolbar. Select **Slope**. Its icon is ◸. Click on each of the four lines to display their slopes.

- Select the icon third from the right in the toolbar. Select **Calculate**. Its icon is 🖩. Click on the slope of one line, for example, $\overleftrightarrow{DC}$.

- In the calculator, click on $\boxed{\ast}$. In the diagram, click on the slope of another line, for example, $\overleftrightarrow{CB}$. In the calculator, click on $\boxed{=}$. This gives the product of the slopes of $\overleftrightarrow{DC}$ and $\overleftrightarrow{CB}$.

- Move the mouse pointer to the answer in the calculator. Depress the mouse button and drag the result to a convenient screen location.

- Rotate $\overleftrightarrow{DC}$ or $\overleftrightarrow{CB}$ until the product of their slopes is –1 to form a right angle.

- Rotate the other two lines until slopes of opposite lines are equal.

Since quadrilateral *ABCD* has one right angle and opposite sides are parallel, quadrilateral *ABCD* is a rectangle.

Circles, Chords, and Similar Triangles

What can be discovered from the placement of four points as shown on a circle and the construction of two triangles also as shown? To find some conclusions, set up an exploration as follows.

- Use the **Circle**, **Point On Object**, **Segment**, and **Intersection Point** tools to draw the circle, points *A*, *B*, *C*, *D*, *E*, and *O* and the six segments that make the two triangles. Label the points as shown.

- Use the **Distance & Length** tool to display the six lengths of the sides of the two triangles formed.

To discover a relationship between △*ABC* and △*DEC*, enter the *Cabri Geometry II™* calculator.

- Select the icon third from the right in the toolbar. Select **Calculate**. Its icon is [🖩]. Click on the length of $\overline{AB}$.

- In the calculator, click on [÷]. In the diagram, click on the length of $\overline{DE}$.

  In the calculator, click on [=]. This gives the ratio of the length of $\overline{AB}$ to that of $\overline{DE}$.

- Place the mouse pointer on the answer in the calculator. Depress the mouse button and drag the result to a convenient location. Release the mouse button and click once.

- Select the icon second from the right in the toolbar. Select **Comments**. In the box that appears, highlight Result: and type AB/DE = instead.

- Proceed in the same way to calculate and display the ratios AC/DC and BC/EC.

The relationship between △*ABC* and △*DEC* is that they are similar.

A second conclusion from the drawing can be seen. If $\dfrac{AC}{DC} = \dfrac{BC}{EC}$, then $AC \cdot EC = BC \cdot DC$. This means the following.

> If two chords in a circle intersect in a point, then the product of the two parts of one chord formed by the intersection point is equal to the product of the two parts of the second chord formed by the intersection point.

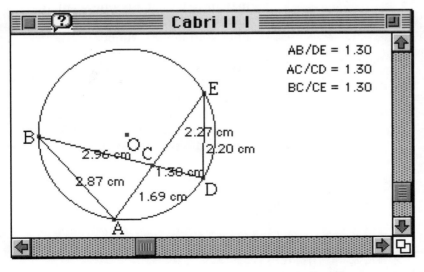

Cabri II I

AB/DE = 1.30
AC/CD = 1.30
BC/CE = 1.30

2.27 cm
2.20 cm
2.96 cm
1.38 cm
2.87 cm
1.69 cm

Frequently, a dynamic drawing in a software program illustrates and makes plausible the truth of a particular statement. Still, the explorer might ask whether it is possible to prove by deductive reasoning that the statement is true, that is, that the statement is a theorem. After completing the exploration of triangles in a circle, attempt to construct a proof using mathematical notation, definitions, and previously proven facts.

Design and Unique Coordinate Systems

A quick glance at the Cabri diagram on this page reveals two facts.

- Two designs produced by a simple algorithm are present.

- Two coordinate systems, neither of which is the standard Cartesian coordinate system, were used along with the design algorithm.

To create the larger part of the design shown, follow these steps.

- Select the icon at the far right in the toolbar. Select **New Axes**. Its icon is

  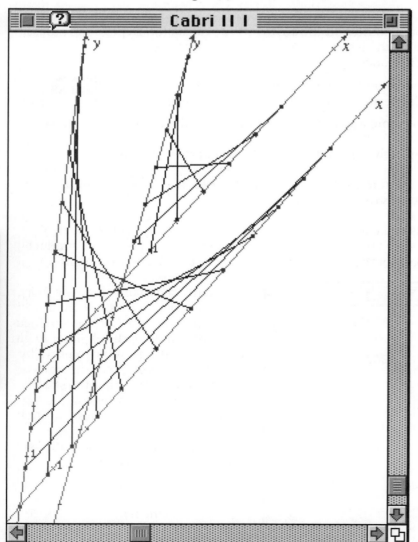
  .

- Click on a point near the lower left of the drawing sheet. Create the axes with their orientation somewhat like those shown here.

- Using the **Segment** tool, click on the 1 tick mark on the left (y) axis. Click on the nth tick mark on the right (x) axis. (In this diagram, $n = 12$.) If, as you bring the pencil pointer close to the tick mark you want, you are asked Which object?, click on These axes.

- Using the **Segment** tool, click on the 1 tick mark on the right (x) axis. Click on the nth tick mark on the left (y) axis.

- Using the **Segment** tool, click on the second tick mark on the left (y) axis. Click on the $(n - 1)$st tick mark on the right (x) axis.

- Using the **Segment** tool, click on the second tick mark on the right (x) axis. Click on the $(n - 1)$st tick mark on the left (y) axis.

- Proceed in this way, according to the algorithm, to join points on the y-axis to points on the x-axis and points on the x-axis to points on the y-axis.

Depending on the size of the drawing sheet and number of tick marks, you may have very many line segments. The more line segments you draw, the finer your design will be.

Algebra students may find this exploration an interesting introduction to parabolas. Geometry students may find this exploration an interesting introduction to different coordinate systems.

The *Geometric superSupposer* Basics

The *Geometric superSupposer* is a well-established dynamic drawing program for computers. This program is both similar to and different from other dynamic drawing programs such as the *Geometer's Sketchpad®* and *Cabri Geometry II™*. That is, you open the program, select various drawing tools to make geometric constructions, measure geometric objects, and, from the on-screen information you obtain, make and consider geometric conjectures.

• To open, or launch, the *Geometric superSupposer*, highlight the *Geometric superSupposer* application name or icon or highlight the name or icon of a particular drawing document. Then double-click on it.

Once you have the *Geometric superSupposer* drawing sheet open, you can begin to create your explorations. The diagram shown here represents one of the principal menu items that you will need when you begin an exploration. If, for example, you want to make a construction whose basic

element is a quadrilateral, click on **Shape**, then **Quad** .

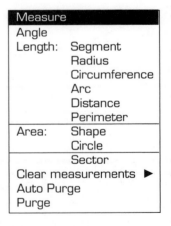

The following two diagrams show two of the major menu items used to construct figures and measure them. When you place the mouse pointer on one of the items followed by ▶, you will see what is called a submenu. For example, if you select **Lines/Triangles ▶** , you will see a short menu that contains the items **Angle Bisector** , **Median** , and **Altitude** .

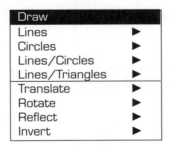

When you have completed your work on a particular exploration, you can save it for future use and then exit the application. Here is how to do that.

• To save a *Geometric superSupposer* document you have created, select **File** , then **Save As...** . Type a name for your document and click on **Save** .

• To close the *Geometric superSupposer* document, click on the small square at the upper left corner of the document.

• To close the *Geometric superSupposer* and open documents, click on **File** , then **Quit** .

On the pages that follow, you will be introduced to various geometry explorations that you can carry out with the aid of the *Geometric superSupposer*.

Quadrilaterals Within Quadrilaterals

What happens if an arbitrary quadrilateral is drawn and the successive midpoints of the sides are located and connected? What can be said about the quadrilateral so formed? Shown in what follows are *The Geometric superSupposer* steps needed to suggest an answer.

- Open *The Geometric superSupposer* and choose **Shape**, then **Quad**. Select any of the options under **▭ Quad**.

- Change the shape and position of quadrilateral *ABCD* by dragging one vertex at a time.

Now locate the midpoint of each side. Here is how to locate the midpoint of $\overline{AB}$.

- Select **Label**, then **Subdivide ▶**, then **Segment m:n**.

- In the dialog box, type AB, 1, and 1. Click on **OK**.

- Proceed in this manner for $\overline{BC}$, $\overline{CD}$, and $\overline{DA}$. Click on **▣▢▦**.

Now draw quadrilateral *EFGH*.

- Select **Draw**, **Lines ▶**, then **Segment**.

- In the resulting dialog box, type EF. Click on **OK**. Proceed in this manner for $\overline{FG}$, $\overline{GH}$, and $\overline{HE}$. Click on **▣▢▦**.

Calculate and record the length of each side of quadrilateral *EFGH*.

- Select **Measure**, then **Length:Segment**.

- In the resulting dialog box, type EF. Click on **OK**. Proceed in this manner for $\overline{FG}$, $\overline{GH}$, and $\overline{HE}$. Click on **▣▢▦**.

To measure an angle of quadrilateral *EFGH*, follow the procedure for finding length except select **Angle** after selecting **Measure**.

These length and angle measures are listed in the Analyze window. Select **Windows**, **Screen mode ▶**, then **Split screen**.

Observe the following.

- Opposite sides of quadrilateral *EFGH* are equal in length.

- Adjacent angles in quadrilateral *EFGH* are supplementary.

Either observation allows you to conclude quadrilateral *EFGH* is a parallelogram.

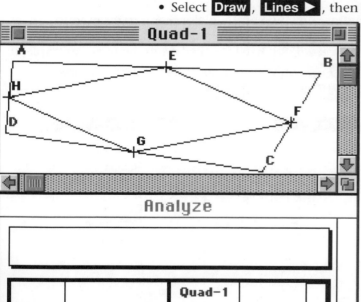

		Quad-1	
1	⊢ HE	6.02	
2	⊢ EF	5.17	
3	⊢ FG	6.02	
4	⊢ GH	5.17	
5	∢ EHG	35.84	
6	∢ HGF	144.16	
7	∢ GFE	35.84	
8	∢ FEH	144.16	

Properties of Reflections

This diagram shows *The Geometric superSupposer* coordinate grid along with a pentagon and its reflection in the vertical axis, $\overleftrightarrow{GF}$. Once the diagram is constructed, questions about properties of reflection can be explored.

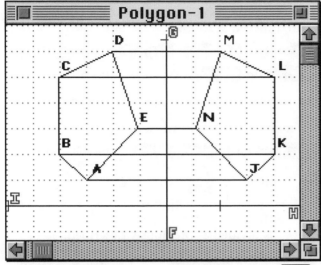

To construct the diagram:

- Select **Edit**, then **Grid** if not already checked.
- Select **Shape**, then **⬠ Polygon**. Under **Polygon**, select **Polygon sides ▶**, and **Pentagon (5)**. Then select **Polygon** and **Random Polygon**.
- Drag the vertices of pentagon *ABCDE* to coordinate locations like those shown.
- Select **Label**, then **Movable point ▶**, then **In plane**. Repeat this process three more times to construct points *F*, *G*, *H*, and *I*. Drag the points to the locations shown.
- Select **Draw**, then **Lines ▶**, then **Segment**.

Type FG. Click on **OK**. Type HI. Click on **OK**, then ▦□▦ .

To construct the reflection of pentagon *ABCDE* across $\overleftrightarrow{GF}$, follow these steps.

- Select **Draw**, then **Reflect ▶**, then **Shape**.
- In the dialog box that appears, type ABCDE, and then GF. Click on **OK**, then ▦□▦ .

To complete the diagram, construct $\overline{DM}$, $\overline{CL}$, $\overline{EN}$, $\overline{BK}$, and $\overline{AJ}$ as previously described.

What information can be gleaned from the diagram?

- The line of reflection, $\overleftrightarrow{GF}$, cuts through the segments joining corresponding vertices at their midpoints.

  To test this observation, locate the midpoint of each such segment by selecting **Label**, then **Subdivide ▶**, then **Segment m:n**. In the dialog box that appears, type the name of each segment whose midpoint you want. Then type 1 and 1. Click on **OK**. Click on ▦□▦ to close the dialog box. You should see that each midpoint is on $\overleftrightarrow{GF}$.

- The line of reflection is perpendicular to the segments joining corresponding vertices.

  To test this observation, find the measures of angles formed by segments joining corresponding vertices and the line of reflection. You should find that they measure 90°.

- Polygons *CDMLKJAB* and *DMNJAE* generated by the reflection process have line symmetry.

  Explore other polygons that are generated by the reflection process and that have line symmetry.

Medians in a Triangle

The medians of a triangle are concurrent in a point. This is a fact from geometry. The point of concurrence is also the center of gravity of the triangular shape if its density is homogenous. A dynamic drawing can bring another fact about medians to light.

To construct the necessary diagram, proceed as follows.

- Select **Shape**, then **▲ Triangle**. Select **Triangle**, **Acute ▶**, and **Scalene**.
- Select **Draw**, **Lines/Triangles▶**, then **Median**. Type ABC and A. Click on **OK**. This will give the median in $\triangle ABC$ from point A. Proceed in this fashion to sketch the other two medians in $\triangle ABC$. Click on ▦.
- Select **Label**, then **Intersection ▶** then **Line with line...** Type AD and BE.

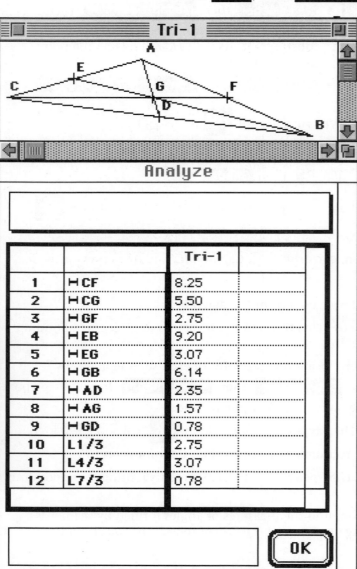

Click on **OK**, then ▦.

- Measure the length of each median and the two segments into which each median is divided as previously described.

- Select **Windows**, then **Screen mode ▶**, then **Split screen** to make the drawing and the Analyze window visible.

To calculate $\frac{2}{3}$ the length of each median, do the following.

- In the Analyze window, place the cursor in the tenth row of the second column.

- Type 2 * L1/3. (L1 refers to the quantity line 1 of this chart, which is the length of median *CF*.)

- Proceed in the same manner to find $\frac{2}{3}$ the length of each other median.

The numbers in rows ten through twelve agree with some of the numbers in rows one through nine.

| In any triangle, the distance between any vertex and the point of concurrence of the medians is two-thirds the length of the median from that vertex. |

Analyze table:

		Tri-1
1	⊢ CF	8.25
2	⊢ CG	5.50
3	⊢ GF	2.75
4	⊢ EB	9.20
5	⊢ EG	3.07
6	⊢ GB	6.14
7	⊢ AD	2.35
8	⊢ AG	1.57
9	⊢ GD	0.78
10	L1/3	2.75
11	L4/3	3.07
12	L7/3	0.78

Circles and Angles

Suppose that you place three arbitrary points *B*, *C*, and *D* on circle *A*. How does the measure of ∠*DBC* compare with that of ∠*DAC*?

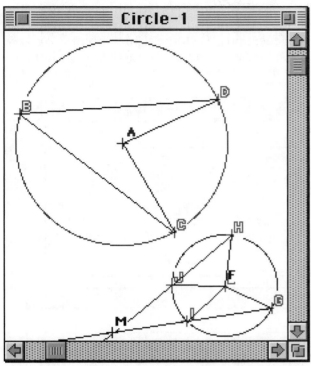

Circle-1

Suppose that you scatter four points *J*, *H*, *G*, and *I* around circle *F* in the order stated. If $\overrightarrow{HJ}$ and $\overrightarrow{GI}$ intersect in point *M*, how does the measure of ∠*HMG* compare with those of ∠*JFI* and ∠*HFG*?

To explore these questions, make a construction in accordance with this general outline.

- Select **Shape**, then **● Circle**. Select **Circle**, then **By radius**. Type a radius. Click on **OK**. Use **Label** and **Movable point ▶** to place three movable points, *B*, *C*, and *D*, on the circle. Construct the line segments shown in this diagram as already described.

- Construct a circle in the plane outside of the first circle. Place four movable points, *J*, *H*, *G*, and *I* on it. Draw the two chords as shown.

- To extend the chords, choose **Draw**, then **Lines ▶**, then **Extend**. (*Note:* When you name the segment to be extended, keep in mind that the extension will begin at the first point you name and pass through the second point you name.) Use **Label**, **Intersection ▶**, **Line with line...**.

- In the second circle, draw in the four radii as shown.

Now you can begin the process of measuring and recording the measures of the angles listed in this table.

Compute the measure of ∠*DBC*. In the **Analyze** window, compute 0.5*the measure of ∠*DAC*. You should find that this product equals the measure of ∠*DBC*. Measure ∠*HMG*, ∠*JFI*, and ∠*HFG*. In the **Analyze** window, compute 0.5*(the measure of ∠*HFG* – the measure of ∠*JFI*). You should find that it equals the measure of ∠*HMG* .

Circle *A*
∠*DBC*
∠*DAC*
Circle *F*
∠*HMG*
∠*JFI*
∠*HFG*

You may wish to add the measures of arcs $\overset{\frown}{CD}$, the arc subtended by the central angle and inscribed angle in the first circle and $\overset{\frown}{JI}$ and $\overset{\frown}{HG}$, the arcs intercepted by secants $\overline{HM}$ and $\overline{GM}$ in the second circle. (*Note:* When you measure a minor arc in *The Geometric superSupposer*, name the endpoints of the arc in counterclockwise order. For example, to measure the length of $\overset{\frown}{JI}$, list *J* first and then *I*. Also note that when you measure an arc, you obtain its length in linear units, not degrees. To measure an arc in degrees, find the length of the arc and the circumference of the circle of which it is a part, then compute

$360*\left(\dfrac{\text{length of arc}}{\text{circumference of circle}}\right)$ in the **Analyze** window.)

Areas of Quadrilaterals

On this page, you will see quadrilateral *ABCD* placed on a square grid. Suppose that you want to find the area of the quadrilateral. You know nothing about the quadrilateral except the positions of the vertices relative to one another. Suppose also that the operation of finding the area of a polygon immediately is not available to you. There are many strategies you could contemplate. Shown here are three illustrated strategies, all of them possible to implement in *the Geometric superSupposer*. You will, however, need to recall and use the geometry formulas for the area of a rectangle, triangle, and trapezoid.

Strategy 1

Place the quadrilateral inside a rectangle. Compute the area of the rectangle by multiplying the lengths of two adjacent sides. Then subtract off the area of each right triangle found by applying the triangle area formula repeatedly.

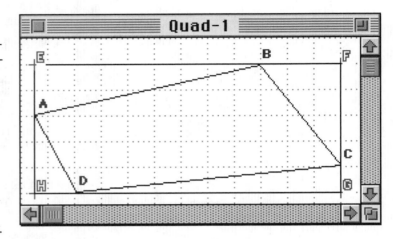

Strategy 2

Draw in a diagonal. Construct the altitudes from two vertices. Use the area of a triangle formula twice to find the areas of the triangles into which the quadrilateral is subdivided. Then add.

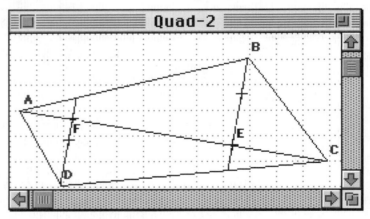

Strategy 3

Subdivide the quadrilateral into a trapezoid and a triangle. Draw in a height for each figure. Apply the area of a trapezoid formula and the area of a triangle formula. Then add.

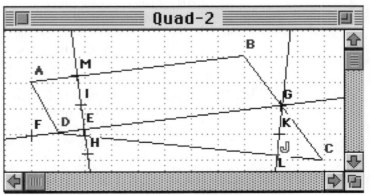

All strategies should give the same answer. You may wish to carry out each strategy using the drawing tools explained in the earlier pages. You might also wish to explore the implementation of other strategies.

A View of the Texas Instruments TI-92

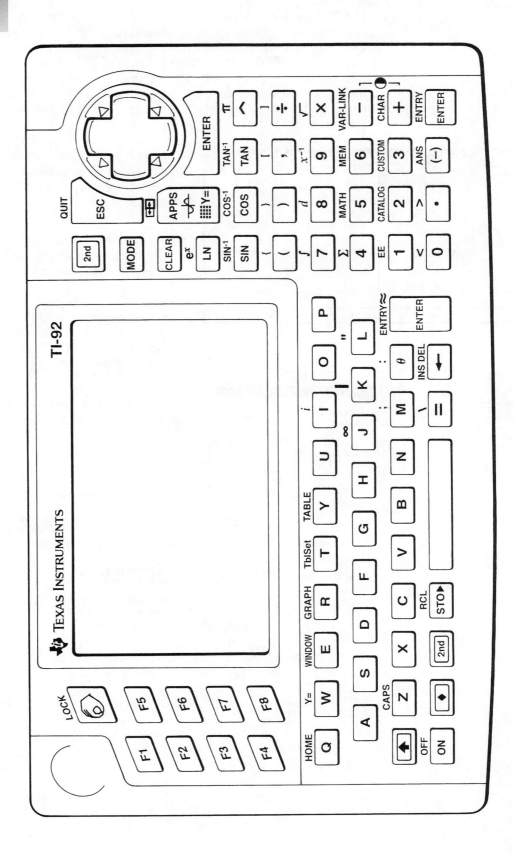

Introduction to the TI-92 in the Classroom

The TI-92 is a hand-held computing device. It resembles both a calculator and a computer. For example, like a calculator, it has a numeric keypad on which you can carry out numerical operations. Like a graphics calculator, it has the capability to graph lists of functions entered at the keypad. However, when you turn on the TI-92, you will notice a menu bar at the top of the display. This menu bar resembles what a computer user would see on a computer screen when a computer application is opened. What follows is a brief tour of some of the menus and submenus.

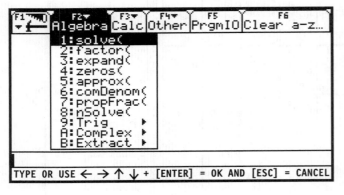

The Algebra menu, or F2 menu, contains a set of algebraic tools. Notice that the last three items in the menu continue. This is indicated by ▶. When you select one of these items, a submenu of choices appears on the display.

- Through the Algebra menu, you can access commands that will factor an expression, expand an expression, solve an equation, and so on.

Also shown on this page is the application menu. To access it, press the APPS button found just to the left of the circular set of arrow buttons. When you press the APPS button and use the cursor pad to choose **8: Geometry ▶**, you will see an interactive display from which you can enter the drawing sheet. The tools enable you to construct points, lines, polygons, midpoints, perpendicular and parallel lines, and so on. In the drawing sheet, you can create geometry explorations.

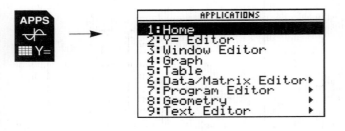

- Through the APPS menu, you can access a large complement of geometry tools.

When you press the APPS button and choose **4: Graph**, you will see an interactive display from which you can enter the graphing sheet.

The selection of APPS and **5: Table** will access the TI-92 spreadsheet.

The selection of APPS and **6: Data/Matrix Editor ▶** will take you into the part of the TI-92 that handles data sets and matrices.

As you become familiar with the TI-92, you may wish to use *keyboard shortcuts*. For example, press these key combinations for options related to graphing.

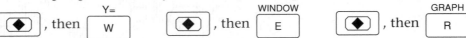

Other useful keyboard shortcuts include the following key combinations.

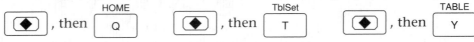

Good luck exploring and using the TI-92. Your ability to work with it will grow the more you use it.

In working with the TI-92, you will need to become accustomed to using many keys and key combinations. This table highlights those you might use most often.

Use the cursor pad to make certain menu selections and to move the cursor from one screen location to another. In the pages that follow, symbols such as ▶ and ▼ will indicate which side of the cursor pad to press.

← Use this key to delete characters immediately to the left of the cursor.

CLEAR Use this key to delete highlighted items.

2nd Use this key in combination with another to access a function, menu, or command printed in orange on the keypad.

ENTER Use this key to execute a command or to say OK to a selection.

ESC Use this key to exit a present menu and return to the menu from which you came.

2nd **ESC** QUIT Use this key combination to exit an application entered from the APPS button and return to the main menu.

◆ Use this key in combination with another key to access functions, commands, and operations printed in green on the keypad.

F1 F2 F3 F4 F5 F6 F7 F8 Use these keys to access particular menus from the home screen, from another menu, or from an application.

MODE Use this key to choose a graph type (function, parametric, polar, sequence, or 3D), a degree of precision in calculations (various FLOAT choices), an angle measurement system (radians or degrees), and so on.

Use this key in combination with the cursor pad to move one or more geometric objects from one screen location to another.

The skills that follow will introduce you to some of the "basics" of the TI-92. For example, Skill 1 shows how to enter an expression like $a^2 + 10a + 25$.

Skill 1: Entering an Expression

Press and type as indicated: a $\boxed{\wedge}$ 2 $\boxed{+}$ 10a $\boxed{+}$ 25 $\boxed{\text{ENTER}}$

Suppose you want to change $a^2 + 10a + 25$ to read $a^2 - 6a + 9$. Skill 2 shows how to either edit the expression or delete it altogether.

Skill 2: Editing and Clearing an Expression

To edit, use a^2+10a+25 in the entry bar at the bottom of the display. Place the cursor to the right of 0. Press $\boxed{←}$ three times to delete +10. Type and press $\boxed{-}$ 6. Place the cursor to the right of 5. Press $\boxed{←}$ twice to delete 25. Type 9. To delete the expression altogether, use the cursor pad to highlight a^2 + 10a + 25 and press $\boxed{\text{CLEAR}}$.

Exploring Polynomial Expressions and Functions

The equation $(x + 2)(x + 3) = x^2 + 5x + 6$ can be interpreted in either of two ways. That is, if you multiply, or *expand*, $(x + 2)(x + 3)$, you get $x^2 + 5x + 6$. Conversely, if you *factor* $x^2 + 5x + 6$, you get the product $(x + 2)(x + 3)$.

The process of multiplying polynomial expressions is called expansion. Suppose you have these four products, all of the same form.

$$(x - 3)(x + 3) \qquad (2x - 1)(2x + 1) \qquad (3x - 2)(3x + 2) \qquad (10x - 11)(10x + 11)$$

Do their expansions have a special form? To find out, use the $\boxed{F2}$, or **Algebra**, menu. Skill 1 shows how to expand $(x - 3)(x + 3)$.

Skill 1: Exploring Special Products

Press $\boxed{F2}$. Use the cursor pad to select ▮**3: expand (**▮, then press and type the

following: $\boxed{\text{ENTER}}$ $\boxed{(}$ $\boxed{x}$ $\boxed{-}$ $\boxed{3}$ $\boxed{)}$ $\boxed{(}$ $\boxed{x}$ $\boxed{+}$ $\boxed{3}$ $\boxed{)}$

$\boxed{)}$ $\boxed{\text{ENTER}}$

Note: As an alternative, you can type expand(directly from the keypad.

Follow this procedure to expand $(2x - 1)(2x + 1)$, $(3x - 2)(3x + 2)$, and $(10x - 11)(10x + 11)$. When all four expansions are displayed, you can see that each consists of a difference of two monomials. The first is the square of the first term in each factor, and the second is the square of the second term.

Now consider the relationship between coefficients b and c of $x^2 + bx + c$ and the numbers r and s in its factorization, $(x + r)(x + s)$. Consider these instances.

$$x^2 - 5x - 6 \qquad x^2 + 8x + 7 \qquad x^2 - 5x + 6 \qquad x^2 + 2x + 1$$

To explore the relationship, you can again use the $\boxed{F2}$, or **Algebra**, menu. Skill 2 shows how to factor $x^2 - 5x - 6$.

Skill 2: Exploring Factoring

Press $\boxed{F2}$. Use the cursor pad to select ▮**2: factor (**▮, then press and type

the following $\boxed{\text{ENTER}}$ $\boxed{x}$ $\boxed{\wedge}$ $\boxed{2}$ $\boxed{-}$ $\boxed{5x}$ $\boxed{-}$ $\boxed{6}$ $\boxed{)}$ $\boxed{\text{ENTER}}$

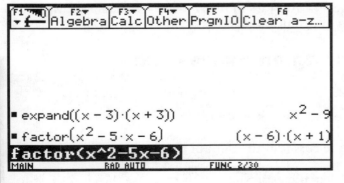

Note: As an alternative, you can type factor(directly from the keypad.

Follow this procedure to factor $x^2 + 8x + 7$, $x^2 - 5x + 6$, and $x^2 + 2x + 1$. When all four factorizations are displayed, you can see that the product of the numbers r and s is c and that their sum is b.

The calculator screen shown here illustrates the display that would appear after the expansion of Skill 1 and the factorization of Skill 2 have been performed.

The TI-92 can be an invaluable aid in exploring the relationship between the factors of a polynomial $P(x)$ and the roots of the polynomial equation $P(x) = 0$. For example, consider the function $P(x) = x^2 - x - 6$.

First factor $x^2 - x - 6$ as described in Skill 2. You will find that the factorization is $(x - 3)(x + 2)$.

Now you are ready to graph $P(x) = x^2 - x - 6$ using the procedure outlined in Skill 3. Note that, to get a reasonable picture of any function, you need to set an appropriate viewing window. For the function $P(x) = x^2 - x - 6$, the ranges $-5 \le x \le 5$ and $-10 \le P(x) \le 10$ will be quite suitable.

Skill 3: Graphing a Function

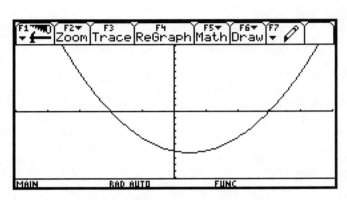

Press ◆, then **Y= W**. Then press and type the following: x **^** 2 **−** x **−** 6 **ENTER**. With the function now entered into the function list, press ◆, then **WINDOW E**. Enter the minimums and maximums for x and $P(x)$. Use ▼ to advance to the next row after a value is entered. Be sure to use **(−)** to indicate a negative value. To display the graph, press ◆, then **GRAPH R**.

The graph shows that the x-intercepts of $P(x) = x^2 - x - 6$ are -2 and 3. Recall that the factors of $x^2 - x - 6$ are $(x - 3)$ and $(x + 2)$. As a result, you might make a conjecture that, if $P(x) = (x + r)(x - s)$, the x-intercepts of its graph are $-r$ and s.

INVESTIGATIONS

1. Try experimenting with the factorization of $P(x) = x^n - 1$ under the conditions given here.

   a. The exponent n is a positive even integer.

   b. The exponent n is a positive odd integer.

2. Suppose that $P(x) = (x - a)(x - b)(x - c)$.

   a. Look for a relationship between the graph and factorization of $y = P(x)$ if a, b, and c are all different. [There are three distinct factors and x-intercepts. One peak and one valley occur between the x-intercepts.]

   b. Look for a relationship between the graph and factorization of $y = P(x)$ if two of a, b, and c are equal and different from the value of the third of a, b, and c. [There are two distinct factors and x-intercepts. One x-intercept is a point of tangency of the graph of $y = P(x)$.]

   c. Look for a relationship between the graph and factorization of $y = P(x)$ if a, b, and c are all the same. [There is one factor that occurs three times. The one x-intercept is a point of inflection for the graph of $y = P(x)$.]

HAND-HELD COMPUTER

Using Tables and Graphs to Explore Patterns

If you deposit P dollars into an account that pays r% interest compounded annually and leave the money untouched for x years, the amount A you will have is given by the following formula.

$$A = P\left(1 + \frac{r}{100}\right)^x$$

Using the TI-92 table and graph features, you can examine how A changes if r takes on some different but fixed values and x is allowed to vary over the set of nonnegative integers. Suppose that you let $r = 4$, 4.5, and 5. Shown in Skill 1 is the procedure for the exploration using a table. For simplicity, let P be one dollar.

Skill 1: Exploring Compound Interest Using a Table

x	y1	y2	y3		
0.	1.	1.	1.		
1.	1.04	1.045	1.05		
2.	1.0816	1.092	1.1025		
3.	1.1249	1.1412	1.1576		
4.	1.1699	1.1925	1.2155		
5.	1.2167	1.2462	1.2763		
6.	1.2653	1.3023	1.3401		
7.	1.3159	1.3609	1.4071		

x=0.
MAIN   RAD AUTO   FUNC

Press ◆, then [Y= W]. Place the cursor to the right of y1=. Press and type the following.

(1 + 4 ÷ 100) ^ x

ENTER (1 + 4.5 ÷ 100)

^ x ENTER (1 + 5 ÷ 100

) ^ x ENTER

Press ◆, then [TblSet T]. Type 0 for tblStart:. Press ▼. Type 1 for Δtbl:.

Check that the other two settings are OFF and AUTO. Press ◆, then [TABLE Y]. The display will be similar to the one shown here.

The table makes it clear that, for a fixed amount of deposit, amount A increases as the rate of interest r increases.

It is easy to use graphing to see the same pattern. Since the compound interest formulas are already entered into the function list, you need only choose a viewing window.

Skill 2: Exploring Compound Interest Using a Graph

Press ◆, then [WINDOW E]. For the minimum value of x, type 0. For the maximum, type 7. For the minimum value of y, type 0. For the maximum, type 1.5. Press ◆, then [GRAPH R].

INVESTIGATIONS

1. **a.** Try investigating the difference between compound interest and simple interest. In the case of simple interest, $A = P(1 + rx)$.

   b. Try experimenting with changing P instead of changing r.

2. Try investigating the volume of a sphere with radius r and how it compares with the volume of a cube with edge of length r.

3. Try looking for a pattern in the family $f_c(x) = x(x - c)$ for different values of c.

Circles and Tangents

The TI-92 diagram shown here contains a circle of radius 1 with center at the origin, a radius, a point of tangency, and a tangent. An interesting fact about the line of tangency can be discerned from the constructions explained here.

Press the APPS button. Select **8: Geometry▶**, then **3: New...** .

Press ENTER . Type a brief variable name, like cirtan. Press ENTER twice.

Press **F8** , select **9: Format...** . Press ENTER . Then make the following choices.

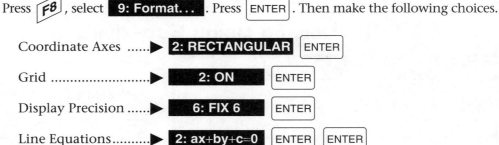

Coordinate Axes ......▶ **2: RECTANGULAR** ENTER

Grid ....▶ **2: ON** ENTER

Display Precision ......▶ **6: FIX 6** ENTER

Line Equations..........▶ **2: ax+by+c=0** ENTER ENTER

Skill 1: Drawing the Diagram

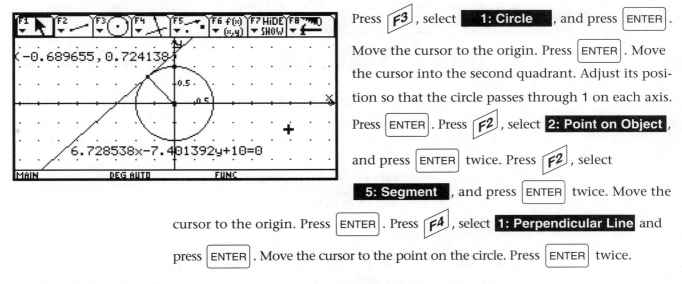

Press **F3** , select **1: Circle** , and press ENTER .

Move the cursor to the origin. Press ENTER . Move the cursor into the second quadrant. Adjust its position so that the circle passes through 1 on each axis.

Press ENTER . Press **F2** , select **2: Point on Object** , and press ENTER twice. Press **F2** , select

5: Segment , and press ENTER twice. Move the

cursor to the origin. Press ENTER . Press **F4** , select **1: Perpendicular Line** and

press ENTER . Move the cursor to the point on the circle. Press ENTER twice.

Skill 2: Displaying Coordinates and Equations

Now display the coordinates of the point on the circle and an equation of the tangent.

Press **F6** , select **5: Equation & Coordinates** and press ENTER twice. Move

the cursor to the line. Press ENTER .

INVESTIGATION

Press **F1** , select **1: Pointer** , and press ENTER . Move the cursor to the

point on the circle. Press and hold ✑ while you use the arrow keys to move

the point around the circle. You should observe that, if (a, b) is on a circle of radius r with center at the origin, then the tangent at (a, b) has equation $ax + by = r^2$.

Solving Systems of Equations

A system of two linear equations in two unknowns is shown here. One way to solve the system is to solve one equation for x in terms of y and then substitute the resulting expression for x in the other equation. Then there is one equation in one unknown, y. When this equation is solved and the value of y is found, that value can be substituted into either original equation. Then that equation is solved for x. Skill 1 illustrates how this strategy is carried out on the TI-92.

$$\begin{cases} 4x + 5y = 13 \\ 3x + y = -4 \end{cases}$$

Skill 1: Solving Using Substitution

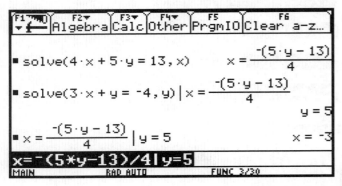

Type and press the following. [◆] [Q] HOME

solve(4x+5y=13,x) [ENTER] solve(3x+y=−4,y)

[2nd] [K]. (Use [(−)] for the negative sign, not [−].) Use ▲ to highlight the equation for x.

Press [F1] and select **5: Copy**. Press [ENTER]. Use ▼ to place the cursor in the entry line.

Press [F1] and select **6: Paste**. Press [ENTER]. Press [ENTER] again for the solution $y = 5$. Use the copy and paste commands from the [F1] menu to put the equation for x in the entry line. Press [2nd] [K]. Use the copy and paste commands from the [F1] menu to put y=5 in the entry line. Press [ENTER]. The display indicates that the solution is $x = -3$ and $y = 5$, or $(-3, 5)$.

A system of three linear equations in three unknowns is shown here in both equation and augmented-matrix form. Skill 2 illustrates a different strategy for solving it. Observe that this strategy involves two matrices, each enclosed in brackets ([ ]). Also notice that each row in each matrix is separated from the others by a semicolon (;).

$$\begin{cases} x - 2y + 3z = 4 \\ 2x + y - 4z = 3 \\ -3x + 4y - z = -2 \end{cases} \longrightarrow \left[\begin{array}{ccc|c} 1 & -2 & 3 & 4 \\ 2 & 1 & -4 & 3 \\ -3 & 4 & -1 & -2 \end{array}\right]$$

Skill 2: Solving Using the Augmented Matrix

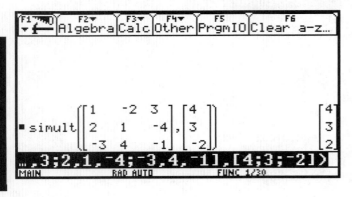

Type simult([1,−2,3;2,1,−4;−3,4,−1],[4;3;−2]).

Press [ENTER].

To access [, press [2nd], then [,].

To access], press [2nd], then [÷].

To access ;, press [2nd], then [M].

So the solution is $x = 4$, $y = 3$, and $z = 2$, or $(4, 3, 2)$.

Of course, the method illustrated in Skill 2 can also be used to solve the system of two linear equations in two unknowns solved in Skill 1.

Since the system of equations solved in Skill 2 contains three equations and three unknowns, a matrix-inverse method may be used to solve the system. This is illustrated in Skill 3.

Keep in mind that the solution can be obtained only if the multiplication of the constant matrix $\begin{bmatrix} 4 \\ 3 \\ -2 \end{bmatrix}$ by the inverse of the coefficient matrix $\begin{bmatrix} 1 & -2 & 3 \\ 2 & 1 & -4 \\ -3 & 4 & -1 \end{bmatrix}$ is done on the left.

Skill 3: Solving Using a Matrix and its Inverse

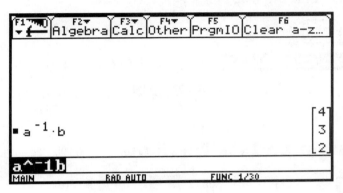

Press the APPS button. Select **6: Data/Matrix Editor**, press ▶, and select **3: New...**. Press ENTER. With the cursor to the right of Type:, press ▶ then select **2: Matrix**. Press ENTER, then press ▼ twice. Type a variable matrix name, such as a. (Be sure to use a name not already in use.) Press ▼. Type the dimensions; in this case, type 3, press ▼, then type 3. Press ENTER twice. In the table that appears, type the entries of the coefficient matrix.

Press ENTER after typing each coefficient. Now press APPS and repeat the steps described above to create the constant matrix. (Give this matrix a name such as b, and enter 3 and 1 as the dimensions.) Press 2nd, then ESC (QUIT). Type and

press a 2nd 9 (X^{-1}) b ENTER in the entry line at the bottom of the display. As was the case in Skill 2, the display will indicate that the solution is $x = 4$, $y = 3$, and $z = 2$, or $(4, 3, 2)$.

INVESTIGATIONS

1. Suppose that you have the following system. $\begin{cases} ax + by = c \\ dx + ey = f \end{cases}$

   a. Try experimenting with different values of a, b, c, d, e, and f to find what solution possibilities there may be.

   b. Try investigating the system $\begin{cases} ax + by = 0 \\ dx + ey = 0 \end{cases}$ for nonzero values a, b, d, and e.

2. Suppose that you have the following system. $\begin{cases} a_{11}x + b_{12}y + c_{13}z = d_{14} \\ a_{21}x + b_{22}y + c_{23}z = d_{24} \\ a_{31}x + b_{32}y + c_{33}z = d_{34} \end{cases}$

   a. Try experimenting with the different methods of solution presented on these pages with a system whose coefficients you choose.

   b. Suppose you have $\begin{cases} a_{11}x + b_{12}y + c_{13}z = d_{14} \\ \qquad\ b_{22}y + c_{23}z = d_{24} \\ \qquad\qquad\ c_{33}z = d_{34} \end{cases}$, where the coefficients of x, y, and z are nonzero. Does the system always have a solution? [Yes]

Medians and the Center of Gravity

The *center of gravity* of an object is the point in space at which it balances. For a triangular region of uniform density, the center of gravity is the point of concurrence of the medians. To explore this concept on the TI-92, you will need to enter the geometry application and display a triangle, its medians, and the coordinates of certain points.

- Begin by pressing the APPS button. In the menu that appears, select **8: Geometry**, press ▶, and select **3: New…** . Press ENTER . Place the cursor to the right of Variable: and type a short name for the exploration, such as medians. Press ENTER twice.

- To place a rectangular coordinate system, press F8 , then select **9: Format…** . Press ENTER . With the cursor to the right of Coordinate Axes…, press ▶, select **2: RECTANGULAR**, and press ENTER . With the cursor to the right of Grid…, press ▶, and select **2: ON** . Press ENTER twice.

Skill 1: Drawing a Triangle and its Medians

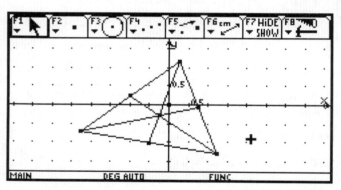

Press F3 , and select **3: Triangle** . Press ENTER . Use the arrow keys to locate the first vertex of the triangle. Press ENTER . Use the arrow keys to locate the second vertex. Press ENTER . Use the arrow keys to locate the third vertex. Press ENTER .

To locate the midpoints of the sides of the triangle, press F4 , select **3: Midpoint** , and press ENTER .

Use the arrow keys to place the cursor on one side. Press ENTER . Repeat this process for the other two sides.

To draw the medians, press F2 , select **5: Segment** , and press ENTER . Place the cursor on one vertex of the triangle and press ENTER . Place the cursor on the midpoint of the opposite side and press ENTER . Repeat this process with the other two vertices.

To mark the point of concurrence of the medians, press F2 , then select **3: Intersection Point** . Press ENTER . Use the arrow keys to place the cursor where the three medians intersect. (The cursor prompt should read POINT AT THIS INTERSECTION.) Press ENTER .

Skill 2: Displaying Coordinates of Points

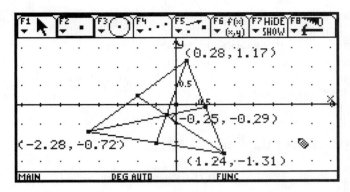

To find a relationship between the center of gravity and the vertices of the triangular region, the next step is to display the coordinates of the vertices and the coordinates of the point where the medians meet. To display the coordinates of a point, press

F6 , select **5: Equation & Coordinates** , and

press ENTER . Use the arrow keys to place the

cursor on a vertex. Press ENTER . Repeat this

process three more times for the other two vertices and the point of concurrence.

Skill 3: Using the Coordinates of the Vertices

The final step is to make calculations that involve the coordinates. Press F6 ,

select **6: Calculate** , and press ENTER . The cursor will now be in the

entry line at the bottom of the display. In the steps that follow, you use ▲ to retrieve a value from the drawing sheet.

Type (. Press ▲, then use the arrow keys to highlight the *x*-coordinate of one

vertex. Press ENTER . Type +. Press ▲, then highlight the *x*-coordinate of a

second vertex. Press ENTER . Type +. Press ▲, then highlight the *x*-coordinate of

the third vertex. Press ENTER . Type)÷ 3 and press ENTER .

You will see the average of the *x*-coordinates of the vertices on the drawing sheet. Repeat this process to calculate the average of the *y*-coordinates of the vertices.

The display now shows that the coordinates of the point of concurrence of the medians, the center of gravity of the triangular region, are approximately equal to the averages of the coordinates of the vertices of the triangle.

To change the triangle to a different one, press F1 , select **1: Pointer** , and

press ENTER . Use the arrow keys to move the cursor to one vertex. Press and hold

🖐 while you use the arrow keys to drag the vertex.

INVESTIGATIONS

1. Try experimenting with different triangles, such as isosceles triangles, isosceles right triangles, or equilateral triangles to see if anything special can be said about the center of gravity.

2. Create a new geometry file containing a square with one vertex at the origin, one side along the positive *x*-axis, and one side along the positive *y*-axis.

   a. Experiment with the meaning of center of gravity for a square.

   b. Change the square to a parallelogram. Experiment with the meaning of center of gravity for a parallelogram.

HAND-HELD COMPUTER

Trigonometric Functions

The function list shown here consists of a parent function, $y_1(x) = \sin x$, and four variations on it. To explore the effect of each variation on the parent, you can graph the functions one at a time. Skill 1 shows how to graph $y_1(x) = \sin x$ and $y_2(x) = \sin (2x)$ over the interval $-360° \le x \le 360°$. Since the interval for x is in degrees, you must set the TI-92 to degree mode.

$$\begin{cases} y_1(x) = \sin x \\ y_2(x) = \sin (2x) \\ y_3(x) = 2 \sin x \\ y_4(x) = \sin (x - 90°) \\ y_5(x) = \sin x + 1 \end{cases}$$

Skill 1: Graphing Trigonometric Functions

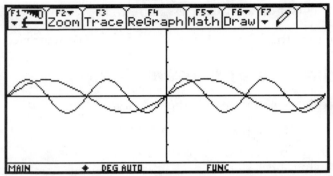

Press [MODE]. With the cursor to the right of Graph..., press ▶, then select **1: FUNCTION**. Press [ENTER]. Use the down arrow key to place the cursor to the right of Angle..., press ▶, then select **2: DEGREE**. Press [ENTER] twice. Now set the viewing window. Press [◆], then [E] (WINDOW). Type the minimum and maximum values of the interval for x. For the minimum and maximum values of y, enter −4 and 4.

Now enter $y_1(x) = \sin x$ and $y_2(x) = \sin (2x)$. To do this, press [◆], then [W] (Y=).

With the cursor to the right of y1=, press and type [SIN] x [ )]. Press [ENTER].

With the cursor to the right of y2=, press and type [SIN] 2x [ )]. Press [ENTER].

Press [◆], then [R] (GRAPH).

The display shows that the graph of $y_2(x) = \sin (2x)$ completes four cycles over the interval $-360° \le x \le 360°$, while the graph of $y_1(x) = \sin x$ completes just two cycles over that interval.

To explore the relationship between the graph of $y_1(x) = \sin x$ and $y_3(x) = 2 \sin x$, press [◆], then [W] (Y=). Place the cursor in the row containing sin (2x) and press [F3]. The cursor will now be in the entry bar at the bottom of the screen.

Place the cursor to the right of *. Press [←] twice to delete 2 and *. Place the cursor to the right of =. Type 2 and press [ENTER]. Press [◆], then [R] (GRAPH). The display will show that the graph of $y_3(x) = 2 \sin x$ extends twice as high and twice as deep as that of $y_1(x) = \sin x$. It has the same period and x-intercepts.

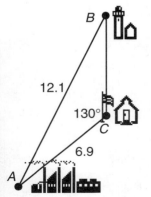

12.1

130°

6.9

The TI-92 can be an invaluable aid in the study of triangle problems. For example, suppose that, in $\triangle ABC$ shown here, $AC = 6.9$ miles, $AB = 12.1$ miles and $\angle ACB$ measures 130°. What is the measure of $\angle ABC$? What will the measure of $\angle ABC$ be if AB is actually 11.9 miles or 12.3 miles?

If, in $\triangle ABC$, $AC = 6.9$ miles, $AB = 12.1$ miles, and $\angle ACB$ measures 130°, then by the law of sines, $\frac{6.9}{\sin B} = \frac{12.1}{\sin 130°}$. Skill 2 shows how to solve this equation.

Skill 2: Solving Triangles Using the Law of Sines

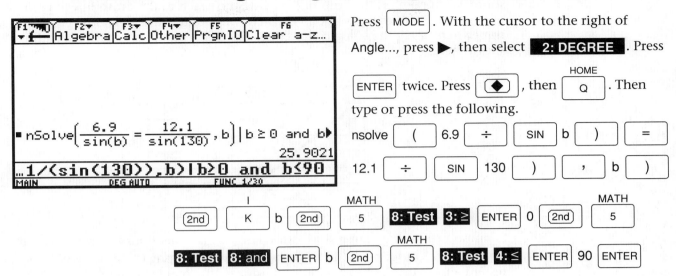

Press [MODE]. With the cursor to the right of Angle..., press ▶, then select **2: DEGREE**. Press [ENTER] twice. Press [◆], then [Q] (HOME). Then type or press the following.

nsolve [( ] [6.9] [÷] [SIN] [b] [ )] [=]

[12.1] [÷] [SIN] [130] [ )] [,] [b] [ )]

[2nd] [|] [K] [b] [2nd] [5] (MATH) **8: Test** **3: ≥** [ENTER] [0] [2nd] [5] (MATH)

8: Test **8: and** [ENTER] [b] [2nd] [5] (MATH) **8: Test** **4: ≤** [ENTER] [90] [ENTER]

The measure of $\angle ABC$, the value of b, will be displayed. It is about 25.90°. Note that the keystrokes include *constraints* on b, since the measure of $\angle ABC$ is between 0° and 90°. The vertical bar is called the *width operator* and is used whenever constraints are desired.

To find the measure of $\angle ABC$ if $AB = 11.9$ miles, you need only edit the current equation in the entry line. To do this, place the cursor in the entry line and to the right of 12.1, press [←] three times, then type 1.9. Press [ENTER]. The new answer will be displayed. Thus, the new measure of $\angle ABC$ is about 26.37°.

You can use the same process to find the measure of $\angle ABC$ if $AB = 12.3$ miles.

INVESTIGATIONS

1. **a.** Try investigating other triangle problems that can be solved using the law of sines. Consider problems in which the measures of two angles and the length of the side between them are given.

   b. Try investigating triangle problems in which the lengths of two sides and the measure of the angle between them are given.

2. This diagram shows a right triangle, $\triangle ABC$, and the lengths of its sides. In $\triangle ABC$, $AB = 10$.

   a. Try experiments in which you find the measure of $\angle ADB$ for different measures of $\angle DAB$ and $\angle DBA$, with point D outside $\triangle ABC$.

   b. Try experiments in which you find the measure of $\angle ADB$ for different measures of $\angle DAB$ and $\angle DBA$, with point D inside $\triangle ABC$.

   c. Do you need to use technology to find the measure of $\angle ADB$ if point D is on $\overline{AB}$? [No; The measure of $\angle ADB$ clearly is 180°.]

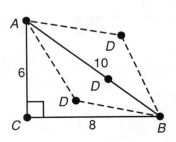

HAND-HELD COMPUTER

Solving a Distance/Time Problem

According to the rules of a competition, contestants begin at a point A, which is offshore and 4.5 miles north of point O. From point A, they row to a point B on shore. From point B, they run or jog along the beach to the finish line at point C, which is 10 miles east of point O. Each contestant is free to choose point B somewhere along the straight shoreline. The graph of one possible race path is shown here.

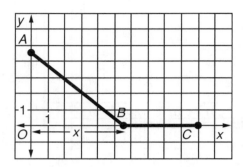

Suppose a contestant can row at a speed of 3.5 miles per hour and run or jog at a speed of 6.5 miles per hour. How far from point O along $\overleftrightarrow{OC}$ should the contestant aim to hit the beach so as to minimize the total race time?

Travel time = $\frac{\text{distance traveled}}{\text{rate}}$, so the function $t(x) = \frac{\sqrt{4.5^2 + x^2}}{3.5} + \frac{10 - x}{6.5}$ models the time it takes to complete the path from A to B to C. One approach to exploring the problem on the TI-92 is to graph this time function. Skill 1 shows how this is done. Note that, since x must be a number in the interval $0 \leq x \leq 10$, you must enter the WINDOW menu and set the minimum of x at 0 and its maximum at 10. Since travel time must also be a nonnegative number, set the minimum of y at 0. Set the maximum at 4.

Be sure that you are in function mode and that previous functions and graphs are cleared.

Skill 1: Exploring the Problem With a Graph

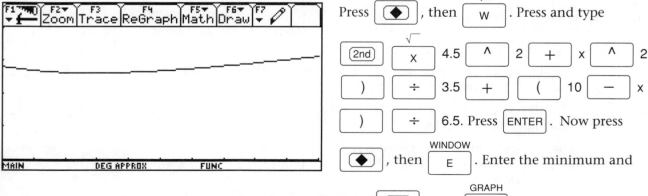

Press ◆, then [W] (Y=). Press and type [2nd] [x] (√) 4.5 [^] 2 [+] x [^] 2 [)] [÷] 3.5 [+] [(] 10 [−] x [)] [÷] 6.5. Press [ENTER]. Now press ◆, then [E] (WINDOW). Enter the minimum and maximum values for x and y. Press ◆, then [R] (GRAPH).

From this graph of the time function, you see that its minimum occurs when the value of x is near 3. Thus, the graph suggests that travel time is minimized when point B is about 3 miles east of O.

You might now want to use a table to get more information about the problem. Since the minimum time results when x is about 3, enter table mode and set the starting value of x at 2. Select 0.1 as the increment value (Δ tbl:) of x. Skill 2 shows how this is done.

Skill 2: Exploring the Problem With a Table

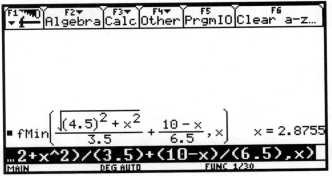

Press $\boxed{\blacklozenge}$, then $\boxed{\text{T}}$ (TblSet). Type 2 and press $\boxed{\text{ENTER}}$. Press ▼. Type 0.1 and press $\boxed{\text{ENTER}}$ twice.

Press $\boxed{\blacklozenge}$, then $\boxed{\text{Y}}$ (TABLE).

Use the arrow keys to scroll through the table. You will find that the minimum value of y_1 in the table is 2.6219. It occurs when $x = 2.9$. So, you have further refined the optimal location of point B to be about 2.9 miles east of point O.

Lastly, you can apply the command that returns a local minimum of a function, as shown in Skill 3.

Skill 3: Exploring the Problem With the Minimum Command

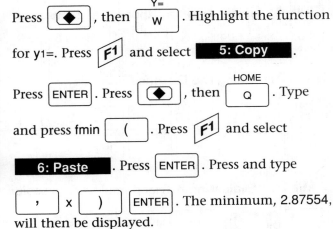

Press $\boxed{\blacklozenge}$, then $\boxed{\text{W}}$ (Y=). Highlight the function for y1=. Press $\boxed{\text{F1}}$ and select ▮5: Copy▮.

Press $\boxed{\text{ENTER}}$. Press $\boxed{\blacklozenge}$, then $\boxed{\text{Q}}$ (HOME). Type and press fmin $\boxed{(}$. Press $\boxed{\text{F1}}$ and select ▮6: Paste▮. Press $\boxed{\text{ENTER}}$. Press and type $\boxed{,}$ x $\boxed{)}$ $\boxed{\text{ENTER}}$. The minimum, 2.87554, will then be displayed.

INVESTIGATIONS

1. Investigate the effect of the varying degrees of precision in the answers when one mile is considered as 5280 feet. [3 miles = 15,840 feet; 2.9 miles = 15,312 feet; 2.87554 miles ≈ 15,183 feet]

2. Try experimenting with the different strategies shown on these pages and variations on the distance/time problem.

HAND-HELD COMPUTER

Evaluating Expressions

Skill 1: Evaluating a Formula From Geometry

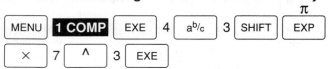

| MENU | 1 COMP | EXE | 4 | aᵇ/c | 3 | SHIFT | EXP (π) |

| × | 7 | ^ | 3 | EXE |

Skill 2: Editing and Reevaluating a Formula

| ◄ | 7 | DEL | SHIFT | DEL (INS) | 8.5 | EXE |

Skill 3: Evaluating a Formula From Algebra

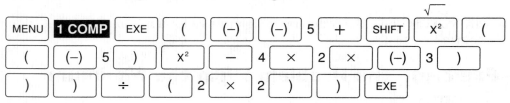

| MENU | 1 COMP | EXE | (| (−) | (−) | 5 | + | SHIFT | x² (√) | (|

| (| (−) 5 |) | x² | − | 4 | × | 2 | × | (−) 3 |) |

|) |) | ÷ | (| 2 | × | 2 |) |) | EXE |

Skill 4: Evaluating an Exponential Expression

| 1250 | × | 1.04 | ^ | (| 1 | aᵇ/c | 3 | aᵇ/c | 4 |) | EXE |

Skill 5: Using the Distance Formula

| x² (√) | (| (| 6 | − | (−) | 2 |) | x² | + | (| 4 | − |

| (−) | 1 |) | x² |) | EXE |

Skill 6: Evaluating a Trigonometric Expression

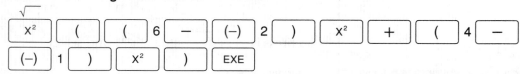

| MENU | 1 COMP | SHIFT | MENU (SET UP) | ▼ | ▼ | ▼ | ▼ | F1 | EXIT |

| 20 | SIN | 32.6 | ÷ | SIN | 47.2 | EXE |

The Graph Viewing Window

Skill 1: Entering Function Graphing Mode

| MENU | 5 GRAPH | EXE | SHIFT | MENU (SET UP) | F1 | EXIT |

Skill 2: Displaying the Current Window Settings

| MENU | 5 GRAPH | EXE | SHIFT | F3 (Window) |

Skill 3: Defining Your Own Viewing Window

Window

| MENU | 5 GRAPH | EXE | SHIFT | F3 | (−) | 6 | EXE | 6 | EXE | 1 | EXE |

| (−) | 4 | EXE | 4 | EXE | 1 | EXE | EXIT |

Skill 4: Choosing a Preset Viewing Window

Window

| SHIFT | F3 | F1 | EXIT |

The Tracing and Zooming Operations

Skill 1: Approximating Coordinates of Points

Trace

| SHIFT | F1 | As you press ▶ or ◀, the coordinates of the selected point

will be displayed.

Skill 2: Zooming in on the Graph

With the graph displayed, place the cursor in the vicinity of a portion of the graph about which you wish to zoom.

Zoom

| SHIFT | F2 | F3 |

Preset Friendly Viewing Windows

Zoom

| MENU | 5 GRAPH | EXE | SHIFT | F2 | F1 | EXIT |

This sets a −6.3, 6.3, 1, −3.1, 3.1, 1 viewing window which yields decimal trace results.

Linear Functions

Skill 1: Graphing a Linear Function

| MENU | 5 GRAPH | Y1: | 1.2 | X,θ,T | + | 2.5 | EXE | F6 |

Skill 2: Graphing a System Linear Functions

| Y1: | 1.2 | X,θ,T | + | 2.5 | EXE | ▼ | Y2: | 1.2 | X,θ,T | − | 1.5 | EXE | F6 |

The color of each function may be changed in the following manner:

| 5 GRAPH | F4 | F1 | OR | F2 | OR | F3 |.

Skill 3: Clearing Functions and Graphs

| Y1= | F2 | F1 | ▼ | Y2= | F2 | F1 |

The Greatest Integer and Absolute Value Functions

Skill 1: Evaluating the Postal Function

[MENU] **1 COMP** [EXE] .32 [+] .23 [OPTN] [F6] [F4] [F2] 7.65 [EXE]

Skill 2: Graphing the Postal Function

Y1: .32 [+] .23 [OPTN] [F5] [F2] [X,θ,T] [EXE] [F6]

To access **ABS**, choose [OPTN] [F5] [F1] .

Systems of Linear Equations

Skill 1: Graphing Linear Equations

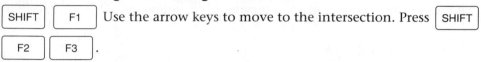

[MENU] **5 GRAPH** [EXE] [SHIFT] [MENU] [F1] [EXIT] **Y1:** 1.4 [X,θ,T] [—]

1.2 [EXE] **Y2:** [( ] 4.59 [—] 1.8 [X,θ,T] [ )] [÷] 2.1 [EXE] [F6]

Skill 2: Tracing and Zooming to the Solution

[SHIFT] [F1] Use the arrow keys to move to the intersection. Press [SHIFT]

[F2] [F3] .

Skill 3: Using the Solver to Solve the System

With the graph displayed, press [SHIFT] [F5 (G-Solv)] [F5] .

Skill 4: Solving a System Algebraically on the Casio Calculators

[MENU] **A EQUA** [EXE] [F1] [F1] [—] 1.4 [EXE] 1 [EXE] [—] 1.2

[EXE] 1.8 [EXE] 2.1 [EXE] 4.59 [F1]

Inequalities

Skill 1: Graphing the Parts of an Inequality

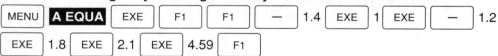

[SHIFT] [MENU] [F1] [EXIT] **Y1:** .8 [X,θ,T] [+] .6 [EXE] **Y2:**

2 [X,θ,T] [X²] [—] 3 [X,θ,T] [—] 1 [EXE] [F6]

Skill 2: Finding the Coordinates of Intersection Points

With the graph displayed, press [SHIFT] [F5 (G-Solv)] [F5] .

Skill 3: Solving an Inequality in Two Variables

[MENU] **5 GRAPH** [EXE] [F3] [F6] [F1] 2 [X,θ,T] [X²] [—]

3 [X,θ,T] [—] 1 [EXE] [F6]

Systems of Linear Inequalities

Skill 1: Graphing Equations in a System

Skill 2: Finding the Coordinates of Intersection Points

With the graph displayed, press | SHIFT | | F5 | | F5 | .

Skill 3: Graphing a Shaded Region

Quadratic Functions: Graphs, Extrema, and Intersection Points

Skill 1: Graphing a Quadratic Function

Skill 2: Finding the Coordinates of a Local Maximum

With the graph displayed, press | SHIFT | | F5 (G-Solv) | | F2 | .

Skill 3: Finding the Coordinates of an Intersection Point

With the graph displayed, press | SHIFT | | F5 (G-Solv) | | F5 | .

Polynomial Functions and Zeros

Skill 1: Exploring the Zeros of a Function

Skill 2: Using a Numerical Method to Find Zeros

The following keystrokes will provide both zeros of the function entered.

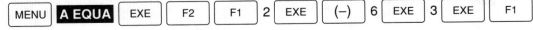

Rational Functions

Skill 1: Graphing a Rational Function

Skill 2 : Graphing a Fuction With a Removable Discontinuity

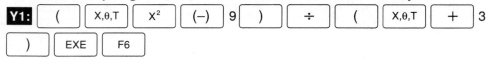

Skill 3: Drawing a Line Segment

To draw a line segment on the graphing window, use the following keystrokes.

With the graph displayed, press SHIFT | SKETCH/F4 | F6 | F2 | F2 . Use the

arrow keys to move the cursor to one end of the line you intend to draw. Press

EXE . Now use the arrow keys to draw the line segment. When finished,

press EXE .

Exponential and Logarithmic Functions

Skill 1: Evaluating the Compound Interest Formula

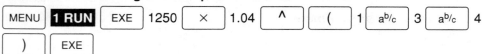

Skill 2: Evaluating Expressions With Radicals

MENU | 1 RUN | EXE | 12.25 | ^ | SHIFT | $\sqrt{}$ X² | 3 | EXE

Skill 3: Evaluating a Special Exponential Expression

3.46 | SHIFT | 10ˣ log | 1.67 | EXE

Skill 4: Solving an Exponential Equation

log | (| 4000 | ÷ | 1250 |) | ÷ | log | 1.04 | EXE

Skill 5: Graphing a Logarithmic Function

Composites and Inverses of Functions

Skill 1: Graphing Functions and Composites

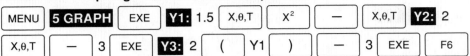

| MENU | 5 GRAPH | EXE | **Y1:** 1.5 | X,θ,T | x² | — | X,θ,T | **Y2:** 2 |

| X,θ,T | — | 3 | EXE | **Y3:** 2 | (| Y1 |) | — | 3 | EXE | F6 |

Trigonometric Functions

Skill 1: Evaluating a Trigonometric Expression

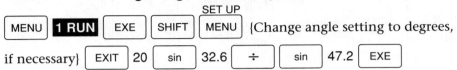

| MENU | 1 RUN | EXE | SHIFT | SET UP MENU | {Change angle setting to degrees,

if necessary} | EXIT | 20 | sin | 32.6 | ÷ | sin | 47.2 | EXE |

Skill 2: Finding an Angle Measure

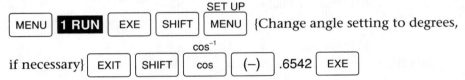

| MENU | 1 RUN | EXE | SHIFT | SET UP MENU | {Change angle setting to degrees,

if necessary} | EXIT | SHIFT | cos⁻¹ cos | (−) | .6542 | EXE |

Skill 3: Graphing a Function Involving Sin

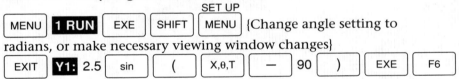

| MENU | 1 RUN | EXE | SHIFT | SET UP MENU | {Change angle setting to

radians, or make necessary viewing window changes}

| EXIT | **Y1:** 2.5 | sin | (| X,θ,T | — | 90 |) | EXE | F6 |

Skill 4: Graphing a Trigonometric Inverse

| **Y1:** 2 | SHIFT | sin⁻¹ sin | X,θ,T | EXE | F6 |

Parametric Equations

Skill 1: Graphing Algebraic Parametric Equations

| MENU | 5 GRAPH | EXE | F3 | F3 | **Xt1:** 2 | X,θ,T | + | 1 | EXE | **Yt1:** |

| 5 | X,θ,T | x² | — | 1 | EXE | F6 |

Skill 2: Graphing Trigonometric Parametric Equations

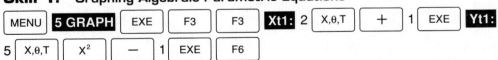

| **Xt1:** 2.5 | cos | (| 1.5 | X,θ,T | + | 2.1 |) | EXE | **Yt1:** 2.4 | sin |

| (| 3.5 | X,θ,T | + | 3 | EXE | F6 |

Skill 3: Graphing a Fuction and its Inverse

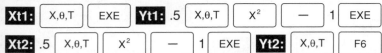

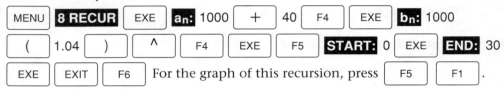

Sequences and Series

Skill 1: Exploring an Explicit Sequence and Skill 2: Exploring a Recursive Sequence

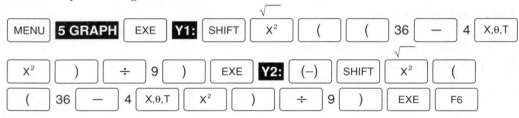

For the graph of this recursion, press F5 F1 .

Second-Degree Equations in Two Variables

Skill 1: Graphing an Ellipse

The Casio 9850 has a built-in Conics section which allows graphing of ellipses, circles, etc… using the exact formula which defines each. This Conics Mode can be accessed by choosing 9 from the menu.

Skill 2: Graphing an Ellipse With Parametric Equations

Entering, Displaying, Editing, and Deleting a Matrix

Skill 1: Entering Matrix A

Skill 2: Displaying Matrix A

Skill 3: Editing Matrix A

Skill 4: Deleting Matrix *A*

MENU | **3 MAT** | F1 | F1

Matrix Operations and Inverses

Skill 1: Entering Matrices *A* and *B*

MENU | **3 MAT** | 2 | EXE | 2 | EXE | EXE | 3 | EXE | 4 | EXE | 1 | EXE | 5 | EXE

EXIT | In the same fashion, select Matrix B, then enter the dimensions and entries for Matrix B.

Skill 2: Finding the Sum of *A* and *B*

$$\overset{A}{}$$

MENU | **1 RUN** | OPTN | F2 | F1 | ALPHA | X,θ,T | + | OPTN

$$\overset{B}{}$$

F2 | F1 | ALPHA | log | EXE

Skill 3: Finding A^{-1}

$$\overset{A}{} \qquad \overset{X^{-1}}{}$$

MENU | **1 RUN** | OPTN | F2 | F1 | ALPHA | X,θ,T | SHIFT |) | EXE

Skill 3: Solving a System of Linear Equations

$$\overset{A}{} \qquad \overset{X^{-1}}{}$$

MENU | 1 | OPTN | F2 | F1 | ALPHA | X,θ,T | SHIFT |) | ×

$$\overset{B}{}$$

OPTN | F2 | F1 | ALPHA | log | EXE

Introduction to One- and Two-Variable Data

Skill 1: Entering One-Variable Data

MENU | **2 STAT** | 6 | EXE | 9 | EXE | 11.4 | EXE | 8 | EXE | 9

Skill 2: Editing a Data Set

MENU | **2 STAT** | (Use  and ⬇ to position the cursor over the 5th data line). 9.5 | EXE | 6.8 | EXE

Skill 3: Sorting the Data

MENU | **2 STAT** | F6 | F1 | EXE | 1 | EXE

Skill 4: Clearing Previously Entered Data

MENU | **2 STAT** | F6 | F4 | F1

Skill 5: Entering Two-Variable Data

MENU 2 STAT 0 EXE 1 EXE 2 EXE 3 EXE 4 EXE 5 EXE ▷

3.5 EXE 3.9 EXE 4.2 EXE 5.8 EXE 6.0 EXE 6.5 EXE

Measures of Central Tendency and Dispersion

Skill 1: Entering the Data

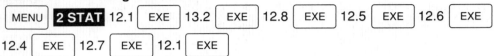

MENU 2 STAT 12.1 EXE 13.2 EXE 12.8 EXE 12.5 EXE 12.6 EXE

12.4 EXE 12.7 EXE 12.1 EXE

Skill 2: Calculating the Mean and Median

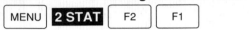

MENU 2 STAT F2 F1

Skill 3: Calculating the Standard Deviation (same as Skill 2)

MENU 2 STAT F2 F1

Histograms

Skill 1: Entering the Data

MENU 2 STAT 4.2 EXE 3.1 EXE ...4.2 EXE

Skill 2: Creating the Histogram

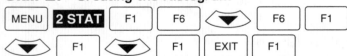

MENU 2 STAT F1 F6 ▽ F6 F1

▽ F1 ▽ F1 EXIT F1

Box-and-Whisker Plots

Skill 1: Entering the Data

MENU 2 STAT 6.4 EXE 7.4 EXE ...7.3 EXE

Skill 2: Creating the Box-and-Whisker Plot

MENU 2 STAT F1 F6 ▽ F6 F2

▽ F1 ▽ F1 EXIT F1

Scatter Plots, Regression Equations, and Correlation Coefficients

Skill 1: Entering the Data

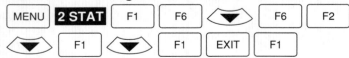

MENU 2 STAT 0 EXE 1 EXE ...5 EXE ▷ 3.5 EXE ...6.5 EXE

Skill 2: Creating the Scatter Plot

Skill 3: Fitting a Linear Model to the Data

View the scatter plot as described in Skill 2. Then press, F1 .

Permutations, Combinations, and Probability

Skill 1: Computing Factorials

Skill 2: Computing Permutations

Skill 3: Computing Combinations

Skill 4: Generating a Random Integer From a Set

Skill 5: Finding Probabilities Involving Combinations

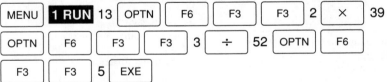

INDEX